Microwave System Engineering Principles

OTHER PERGAMON TITLES OF INTEREST

Baden Fuller – *Worked Examples in Engineering Field Theory*

Chen – *Theory and Design of Broadband Matching Networks*

Coekin – *High Speed Pulse Techniques*

Fisher & Gatland – *Electronics: From Theory into Practice (2nd Ed)*

Garnell – *Guided Weapon Control Systems*

Gatland – *Electronic Engineering Application of Two Port Networks*

Smith – *Efficient Electricity Use*

Microwave System Engineering Principles

by

Samuel J. Raff, Ph.D.
Program Manager U.S. National Science Foundation
Adjunct Professor of Electrical Engineering
George Washington University

PERGAMON PRESS
Oxford / New York / Toronto / Sydney / Paris / Frankfurt

Pergamon Press Offices:

U.S.A.	Pergamon Press Inc., Maxwell House, Fairview Park, Elmsford, New York 10523, U.S.A.
U.K.	Pergamon Press Ltd., Headington Hill Hall, Oxford OX3, OBW, England
CANADA	Pergamon of Canada, Ltd., 207 Queen's Quay West, Toronto 1, Canada
AUSTRALIA	Pergamon Press (Aust) Pty. Ltd., 19a Boundary Street, Rushcutters Bay, N.S.W. 2011, Australia
FRANCE	Pergamon Press SARL, 24 rue des Ecoles, 75240 Paris, Cedex 05, France
WEST GERMANY	Pergamon Press GmbH, 6242 Kronberg/Taunus, Frankfurt-am-Main, West Germany

Library of Congress Cataloging in Publication Data

Raff, Samuel J
Microwave system engineering principles.

Bibliography: p.
Includes index.
1. Microwaves. I. Title.
TK7876.R37 1977 621.381'3 76-52389
ISBN 0-08-021797-4

Printed in the United States of America

To Anna Raff, without whose patience this book would not have been completed and

to Lillian Raff, without whose patience this book would not have been begun.

PREFACE

This is a book of nuggets, principles and the "big picture". Most scientists and engineers working in the field of microwaves are specialists in information theory, antennas, circuits, propagation, detection, reliability, or some equally localized field. They know their field of specialization in much greater depth than the treatment given here. It is the coverage of subjects outside their specialty which is intended to be of primary value to them, in broadening their horizons and providing the key principles and relationships which guide the disciplines which interface with theirs. However, the sections dealing with their specialty may also be of value because the approach is generally different from the usual one, and the diversity in point of view may provide fresh insights.

It is assumed that the reader is familiar with the calculus, differential equations and transforms, although the presentation does not lean heavily on the latter two subjects. I have tried to develop the key formulae and principles with a minimum of mathematics from concepts which are intuitively obvious. When a choice had to be made between rigor and clarity, the latter has been chosen.

The numbers, formulae and relationships which are of adequate significance to the field to warrent memorizing or ready reference are indicated by the symbol "m" in the right hand margin and listed in appendix A. The reader will surely find that he already knows many of them.

The book was compiled from notes developed during eight years of teaching a graduate course on the subject and was used as a text. Thus it has been student tested. The student must have some degree of scientific and engineering maturity in order to grasp the wide range of material. In particular he must be familiar with much of the subject matter by virtue of specialized courses which he has taken. On the other hand, much of the material was new to each student. He had to work harder at these parts. The problems in appendix F were generated when the material was being used as a text, and they may well be useful to the specialist as well as the student.

I trust that readers will find the majority of this book to be of value in clarifying and solidifying their understanding of subjects which with they are somewhat familiar. Surely some parts will be new and some thoroughly familiar to each reader. In striving for lucidity I may have explained some subjects at a level so elementary as to require an apology to those who are already familiar with the particular subject. I apologize.

Samuel J. Raff

Chevy Chase, Md.

CONTENTS

CHAPTER I

THERMAL NOISE

1.1 Basic Nature and Principles Which Can Be Derived From It.

The basic limitation of microwave systems is thermal noise. This is sometimes called Johnson noise or Nyquist noise after two of the principal researchers in this field. J.B. Johnson and H. Nyquist published the two definitive articles on this subject in 1928 (Ref.1,2). It is surprising that our understanding of thermal noise is so recent. Consider that Maxwell developed the theory of the electromagnetic field in the 1870's and the Quantum theory and relativity date from the early 1900's. Thermal noise is quite as basic a phenomenon as these others and more easily understood. It is noted in References 1 and 2 that thermal noise is closely related to Brownian movement and its magnitude is derived from thermodynamics by considering the electrons in a conductor as particles which, like any other particle, have kinetic energy 1/2 kT per degree of freedom. (k is Boltzman's constant. See Eq.1.1) In fact, thermal noise is a modified form of the tail of the black body distribution curve. However, in what follows we shall take a more empirical approach.

Suppose we were to take a resistor into a perfectly shielded room and measure the average voltage across it with a high impedance voltmeter that had at its input a very sharp skirted filter of bandwidth B. Thermal noise theory predicts that the measurement would give a root mean square average voltage.

$$V_{r.m.s.} = \sqrt{4RkTB} \qquad (1.1)$$ m

where $V_{r.m.s.}$ = root mean square average voltage measured (volts)

R = resistance (ohms)

k = Boltzmann's constant (1.38×10^{-23} watt-second/degree) m

T = absolute temperature (degrees Kelvin)

B = bandwidth of the sharp skirted filter (Hertz)

Notice that the noise per unit bandwidth is independent of frequency. This is a characteristic of white noise. However, no real resistor truly puts out white noise over a very large bandwidth because the self inductance of the resistor and the capacitance between its ends will reduce its output at high frequency.

To get some idea of the magnitude of this voltage, assume that R is 1000

---

m Those equations and numbers which, in the opinion of the author are central to the subject and worth memorizing are marked with a "m". They are listed in appendix A.

ohms, T is 290° (this is nominal room temperature), and B is 1 megahertz. Substitution of these values gives an r.m.s. voltage of 4 microvolts, which is small but quite significant.

We can also consider thermal noise from the point of view of available noise power. Consider that a load is put across that resistor. The voltage of Eq.1.1 will cause currents to flow in that load and hence some power will be delivered to it. It is easy to show that the maximum power transfer will occur if the load presents no reactance and a resistance, equal to that of the resistor which is putting out the thermal noise. In this case, the voltage across the load will be half of that which we have calculated above (due to the voltage drop in the resistor which is generating the thermal noise) and one can readily show that the power delivered will be kTB.

Thus, the available thermal noise power from any resistor is kTB. At room temperature this is 4×10^{-21} watts per Hz or -204 dBW. This is an extremely simple expression and one might tend to think of it as some sort of a fictional or imaginary power which is available. However, this is not the case. There is a real power transfer. Of course, if both resistors are at the same temperature, an equivalent amount of power is transferred in the other direction so that the net power transfer is 0. If, however, the resistors are at different temperatures, the net transfer of power is not 0, and the colder resistor will be warmed by power transferred from the warmer resistor. This is quite independent of any thermal conduction along the wires connecting the resistors or of any radiated heat transfer between them.

If there is inductance or capacitance in the resistor which is generating the noise voltage, that does not affect the available noise power at any selected frequency since matching the load merely requires that there be equivalent capacitance or inductance (equal and opposite reactance) in the load to achieve resonance and transfer the same power. This is the normal definition for a load matched to a reactive generator.

Notice also that the available noise power is *independent of the type of material* of which the resistor is made. This is an essential characteristic if we are not to violate the second law of thermodynamics. It can be explained in the following way. Suppose that the value of k were different for different materials. Let us suppose that it is larger for copper than for graphite. That would mean that a copper resistor and a graphite resistor at the same temperature would have different available noise powers. Let us take a copper resistor and a graphite resistor having the same resistance value and at the same temperature and connect them together. If they had different available noise powers one would be delivering more electrical power to the other than was being returned. The net effect would have to be that the graphite resistor would be heated up because it received more power than it delivered. The arrangement would stabilize with the graphite resistor at a slightly higher temperature than the copper resistor. The second law of thermodynamics, which is usually expressed in the form of entropy change, can be interpreted as saying that the net flow of energy between two bodies on which work is not being done, is from the warmer body to the colder body; with a correlary that two bodies at the same temperature which are not acted upon by any third body or by work or by chemical energy, will remain at the same temperature. If my original supposition were correct, this law would be violated. Thus, if two resistors, initially at the same temperature, develop a temperature difference between them by virtue of merely being connected together, they would violate the second law of thermodynamics.

This can be expressed in terms of perpetual motion machines. Thermodynamics tells us that any time we have two bodies at different temperatures we can run a heat engine between them and do work, using the hotter body as the upper temperature heat reservoir and the colder body as the lower temperature reservoir. If there is some way to maintain this temperature difference between the reservoirs, we can run the heat engine indefinitely. If we have two materials of which we can make resistors which have different available noise powers at the same temperature, we can connect them together and they will develop a difference in temperature between them. We can now run a heat engine between them and part of the temperature difference will continue to be maintained by the difference in thermal noise power. Thus we will have a perpetual motion machine. Perhaps the output will be small, but still we will perpetually generate power or motion. However, perpetual motion machines cannot be. Therefore, clearly the available noise power must not depend on the material and must be a universal property.

We can go a little further with this line of reasoning by looking at conditions when the load is not matched to the noise generating resistor. We will assume that load and generator are purely resistive. The extension to reactive components is obvious. In this case the voltage across the load resistor is given by Eq.1.2 and the power delivered to it by Eq.1.3.

$$V_L = V_n \frac{R_L}{R_L + R} \tag{1.2}$$

$$P_L = V_n^2 \frac{R_L}{(R_L + R)^2} = 4kTB \frac{RR_L}{(R + R_L)^2} \tag{1.3}$$

where V_n = r.m.s. noise voltage (from Eq.1.1)

R_L = load resistance

R = noise generating resistance.

One can readily see by the symmetry in Eq.1.3 that the power delivered *by the load to the noise generating resistor* is given by exactly the same expression as Eq.1.3 except, of course, that the temperature will be the temperature of the load. Thus, if the two are at the same temperature, the net exchange of power will be 0, regardless of the resistances involved. One may view this as being fortunate for the second law of thermodynamics because otherwise we could run a perpetual motion machine between two resistors if they were of different value. Alternatively, of course, the second law of thermodynamics is inviolable, and the fact that the relation comes out so neatly indicates that we have the proper form of the equation for the noise voltage. Specifically, the r.m.s. noise voltage must be proportional to the square root of the resistance.

Next, consider the bandwidth B. We have assumed that there is some type of a filter between the two resistors. We have further assumed, although we haven't said it, that this is in fact a lossless filter. All the energy which is not transmitted through it is reflected. How do we know, however, that the bandpass of such a filter is the same looking in one end as it is looking in the other end? The easiest way to know it, of course, is to have faith in the second law of thermodynamics. If the bandwidth of the filter were different looking in one direction from what it is looking in the other, there would be a net exchange of power between the two resistors at the same temperature. Since this is not possible, we have, in fact, derived a property of lossless filters; i.e., *the transmission bandwidth must be the same in both directions.*

As you might suppose, it is not essential that the filter be sharp skirted, and the property can be expressed in a more general way for lossless filters by Eq.1.4.

$$\int_0^\infty |H_{12}(f)|^2 \, df = \int_0^\infty |H_{21}(f)|^2 \, df \tag{1.4}$$

where f = frequency

$H_{12}(f)$ = filter transfer admittance from 1 to 2, i.e., the ratio of output current to input voltage

$H_{21}(f)$ = filter transfer admittance from 2 to 1.

The vertical bars indicate the magnitude of the quantity inside.

One can go even further with this concept by considering that in principal at least, we could insert an additional lossless filter in the circuit which is as narrow as we please. The effect of such an additional filter will be not to pass, in either direction, signals at frequencies outside its passband. The second law of thermodynamics will now be violated if Eq.1.4 does not hold over this new, very restricted, band of frequencies. In the limit of narrowness of this additional filter, the second law of thermodynamics requires that

$$|H_{12}(f)| = |H_{21}(f)| \tag{1.5}$$

Thus the second law of thermodynamics requires, for any lossless filter having the same input and output impedance, that the forward and reverse transfer admittance be identical in magnitude at any frequency. Notice that although this result was derived from the concept of thermal noise, if the circuits are linear it is a very general result about the characteristics of lossless filters.

One can derive another general result, about mismatches in transmission lines, by exactly the same line of reasoning. This is that the reflection coefficient of any lossless transmission line and circuit combination at a particular frequency must be the same when viewed in one direction as when viewed in the opposite direction. Or, consider any two port network having an input terminal and an output terminal; if the circuits are lossless, the reflection coefficient seen looking into the input terminal at a particular frequency must be the same as that seen looking into the output terminal at the same frequency. This, of course, is also a general result about circuits and transmission lines, one which is a necessary property of all lossless circuits if the second law of thermodynamics is not to be violated by the inequality of noise power exchange between two resistors at the same temperature.

This would seem to imply that all lossless circuits and networks must be reciprocal. That is, if there is a low reflection coefficient in going from A to B, there must be the same reflection coefficient in going from B to A. This is true, but only for lossless networks. Where losses are involved the situation is much more complicated because one must consider the thermal noise power being put out by the resistive elements in the network; and under these circumstances it is difficult to draw useful conclusions from noise and thermodynamics about the properties of the networks. However, the conclusions which we have reached are quite valid for lossless networks and they apply to any level of power which might be put into the network at any frequency.#

There is one further principle which can be derived directly from this basic concept of thermal noise and the second law of thermodynamics. This is Thevenin's Theorum or the reciprocity theorum. Consider two points A and B in a circuit. Let me open the circuit at A, insert a zero impedance generator, and measure the current at B due to it. I designate this current as $I_B(\langle V_A \rangle)$. For a linear circuit there is some quantity Q_1 such that

$$I_B\ (\langle V_A \rangle) = Q_1\ V_A \tag{1.6}$$

Next if I open the circuit at B and insert a zero impedance generator at that point, the current at A due to a voltage at B will be similarly designated as

$$I_A\ (\langle V_B \rangle) = Q_2\ V_B \tag{1.7}$$

Thevenin's reciprocity theorum says that the constants of proportionality are the same, i.e., $Q_1 = Q_2 = Q$. This indicates a particular type of reciprocal relationship between any two points in a circuit; and note that it is a relationship between current and voltage of the general type which we designate as a transfer admittance. It is thus a generalization of Eq.1.5.

Let us see what Thevenin's Theorum means in terms of noise power transfer. If we have resistors at A and B, and they are each treated as contain-

---

#This result is derived from the unitary properties of a two port scattering matrix for a lossless junction in appendix C.

ing a noise generator, then the power delivered to the resistor at B due to the voltage of the noise generator at A is given by

$$P_B\ (<V_A>) = Q^2\ V_A{}^2\ R_B = 4kT_A B\ Q^2\ R_A\ R_B \tag{1.8}$$

and similarly

$$P_A\ (<V_B>) = Q^2\ V_B{}^2\ R_A = 4kT_B B\ Q^2\ R_A\ R_B \tag{1.9}$$

Equations 1.8 and 1.9 are, of course, equal if the temperatures are the same. Thus Thevenin's Law assures that there is no net transfer of thermal noise power between two resistors at the same temperature any place in a passive circuit.

There are other laws like Thevenin's Theorum which one could write which do not assure zero net transfer of power. These laws are wrong. For example an analog of Eq.1.6 is

$$V_B\ (<V_A>) = K_1\ V_A \tag{1.10}$$

Using this Eq. the power transferred from a resistor at A to one at B can be written as

$$P_B\ (<V_A>) = K_1^2\ V_A{}^2/R_B = 4kT_A B\ K_1^2\ R_A/R_B \tag{1.11}$$

Similarly,

$$V_A\ (<V_B>) = K_2\ V_B$$

and

$$P_A\ (<V_B>) = k_2^2\ V_B{}^2/R_A = 4k\ T_B B\ K_2^2\ R_B/R_A \tag{1.12}$$

If K_1 was always equal to K_2 then with both resistors at the same temperature,

$$P_A\ (<V_B>) \neq P_B\ (<V_A>) \text{ for } R_A \neq R_B \tag{1.13}$$

and the second law of thermodynamics would be violated because a large resistor would transfer more power to a small resistor than the small resistor would return to it. Thus if all the circuit elements started out at the same temperature, the smaller resistors would heat up while the larger resistors cooled down until equilibrium set in with each resistor at a different temperature. This is clearly impossible in a passive circuit, so $K_1 \neq K_2$ and what looks like an equivalent of the reciprocity theorum is wrong.

On the other hand one can go through steps similar to Eq.1.6 through 1.9 using a transfer impedance instead of an admittance and find no contradiction to the second law of thermodynamics. In fact, Thevenin's reciprocity theorum is equally valid for transfer admittances or impedances, but it is not valid for relations between voltage and voltage, or current and current.

One additional computation will be made to demonstrate the basic nature of thermal noise. This is a calculation of the energy stored in a capacitor with a resistor connected across it. The general circuit is shown in Fig.1.1.

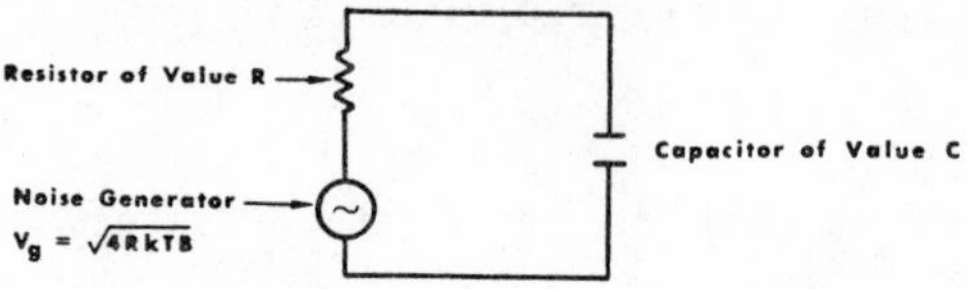

Fig. 1.1. Schematic for calculating noise voltage across capacitor.

If we designate the voltage across the generator shown in FIg.1.1 by the subscript g. and the voltage across the capacitor by the subscript c, we can write

$$|I(f)| = \frac{|V_g(f)|}{\sqrt{R^2 + \dfrac{1}{\omega^2 C^2}}} \tag{1.14}$$

where I(f) is the current flowing in the circuit and $\omega=2\pi f$

$$|V_c(f)| = \frac{|I(f)|}{\omega C} \tag{1.15}$$

Using the fact that the time average of the product of two sinusoids of different frequency is zero, the average energy stored in the capacitor can be expressed as

$$\overline{E} = \frac{1}{2} C\, V_C^2 = \frac{1}{2} C \int_0^\infty |V_c(f)|^2\, df \tag{1.16}$$

where $|V_c(f)|^2$ is the mean square voltage per Hz of bandwidth.

(Note that the factor 1/2 is correct even though V_c is the r.m.s. average value of the voltage.) Substituting Eq.1.15, 1.16 and 1.1 gives

$$\overline{E} = \frac{1}{2} C \int_0^\infty \left|\frac{I(f)}{\omega C}\right|^2 df = \frac{1}{2} C \int_0^\infty \frac{4RkT\, df}{1 + R^2\omega^2 C^2} \tag{1.17}$$

If we express df as $d\omega/2\pi$

$$\overline{E} = \frac{kT}{\pi} \int_0^\infty \frac{R\, C\, d\omega}{1 + R^2\omega^2 C^2} = \frac{kT}{\pi} \int_0^\infty \frac{dx}{1 + x^2} \tag{1.18}$$

where $x = RC\omega$. We can identify the integral as an arc tangent. Whence

$$\bar{E} = \frac{kT}{\pi} \tan^{-1} x\Big]_0^\infty = \frac{1}{2} kT \qquad (1.19)$$

The result indicated in Eq.1.19 is familiar from modern physics. It says that the energy stored in the capacitor is, on the average, kT/2 where k is Boltzmann's Constant. From modern physics we know that any mode of vibration or oscillation which is capable of storing energy will have the zero point average stored energy kT/2. Thus the result is very basic. Notice that this average stored energy in the capacitor is quite independent of the resistance across it and hence would be the same even if that resistance were infinite or zero.

However, at this point there is a small problem to resolve. We started with the thermal noise characteristics of a resistor and showed that the average stored energy in a capacitor is kT/2 if a resistor is connected across that capacitor. However, what if the temperature of the capacitor and the temperature of the resistor are different? The T in thermal noise output of the resistor is clearly the temperature of the resistor, while the T in the kT/2 stored energy of the capacitor (from modern physics) looks like it should be the temperature of the capacitor. The problem could be readily resolved by actually doing an experiment in which the resistor and the capacitor were separated and held at different temperatures. If this experiment is done, one will find that the T in both belongs to the resistor if the resistor is of small value. However, if no resistor is connected across the capacitor, or if the resistance is so large that the dielectric losses in the capacitor dominate, then, of course, it is the temperature of the capacitor which will determine its stored energy.

1.2 Noise Bandwidth, Noise Figure, and Noise Temperature.

Before proceeding with our main subject, it is necessary to make an accurate definition of noise bandwidth, the quantity which we have designated B in Eq.1.1 above. In that section we avoided the problem by referring to filters which are square and have very steep slopes, for which case the bandwidth is the frequency difference between the two cutoffs. In the more general case the noise bandwidth, sometimes called the effective noise bandwidth of an amplifier or filter, is given by Eq.1.20.

$$B_n = \frac{\int_0^\infty |H(f)|^2 \, df}{|H(f_0)|^2} \qquad (1.20)$$

where $H(f)$ = frequency response characteristic of the filter, amplifier or other device in the line. (Transfer impedance, or admittance or voltage or current transfer function.)

f_0 = frequency at which the gain is measured. (Usually the frequency of maximum response).

The logic of this definition is clear since, if we multiply the numerator by the proper constant it will represent the noise transmitted through the amplifier from a white noise source. The denominator represents the power gain of the amplifier. Consider applying Eq.1.20, for example, to a flat top steep sided amplifier with power gain G, bandwidth B and the same input and output impedance. The integral in the numerator will be simply GB and the denominator will be G. Thus in that case B_n is the same bandwidth B we used in Eq.1.1.

The concept of *noise figure* is a simple one but one which is sometimes incorrectly oversimplified. It is a property of an amplifier, mixer, attenuator, filter or other component which relates the output noise to the input noise under special circumstances. To measure the noise figure of an amplifier for example, we put a *matched resistor at room temperature T_0* at its input and we measure its output noise power. We also measure its gain and its noise bandwidth. The noise figure is then given simply by

$$F = \frac{\text{Noise Power Out}}{k\ T_0\ B_n\ G} \qquad (1.21)$$

where G = power gain of the amplifier at its frequency of maximum response

T_0 = room temperature (290°)

and the other symbols have been previously defined.

Recognizing that G is the ratio of signal power out to signal power in, we can write an alternative definition of F by substituting S_{out}/S_{in} for G in Eq.1.21. With this substitution we can show that

$$F = \frac{S_{in}\ N_{out}}{N_{in}\ S_{out}} = \frac{(S/N)_{in}}{(S/N)_{out}} \quad (\text{if } N_{in} = kT_0\ B_n) \qquad (1.22)$$

Thus if we had a perfect amplifier which merely amplified the input noise without adding any noise of its own, its noise figure would be one. If we connect a room temperature matched resistor at the input of a real amplifier with noise figure F, its output noise would be simply F times what we would expect from a perfect amplifier. This makes it a very convenient quantity to use. F may be expressed as a ratio or in dB.

However, *if the resistor at the input of the amplifier is not at temperature T_0,* the noise out of the amplifier is not the same as it would be if the resistor were at temperature T_0 nor is it F times what one would expect from a perfect amplifier. In this case *the simple use of the noise figure F is a mistake.* The output noise must be calculated using effective noise temperatures as described below.

Since most of the circuits with which we deal in microwave communication are linear or at least are described by linear approximations, it is clear that the output noise from an amplifier really consists of two parts, one of which is the input noise amplified, and the other is noise added by the amplifier itself. While an amplifier or other component can be described in terms of its noise figure, the physical reality is that the amplifier adds a certain amount of noise to the input noise or input signal which is substantially independent of the input signal or input noise itself. Thus when the input noise changes, only one of the two parts of the output noise of the amplifier will change; that is, the amplified input noise. The noise added by the amplifier will remain the same. This noise which is added by the amplifier is described in terms of *effective noise temperature.*

Thus we can write for an amplifier with a matched resistor at its input

$$\text{Noise out} = (k\ T\ B_n + \Delta N)\ G \qquad (1.23)$$

where ΔN = measure of the noise aided by the amplifier.

However, in general, this noise may be added at different stages in the amplifier and in different ways. We must measure that noise, of course, at the output; however, ΔN, in Eq.1.23 refers that noise back to the amplifier input. This is indicated by the fact that it is multiplied by G.

We can express this added noise, ΔN, in terms of an effective noise temperature by simply writing

$$\Delta N = k \, T_e \, B_n \tag{1.24}$$

where T_e = effective noise temperature of the amplifier.

Notice that T_e is a completely fictional temperature because ΔN is not necessarily thermal noise. For example, it may be power supply ripple or harmonics of it which occur in the bandwidth of the amplifier. However, it is a very useful concept because if we substitute Eq.1.24 in Eq.1.23 we have

$$\text{noise out} = k \, B_n \, G \, (T + T_e). \tag{1.25}$$

If the input noise resistor happens to be at temperature T_0, then Eq.1.21 applies and we can combine it with Eq.1.25 to obtain:

$$\text{noise out} = F \, k \, T_0 \, B_n \, G = k \, B_n \, G \, (T_0 + T_e) \tag{1.26}$$

or

$$F = 1 + \frac{T_e}{T_0} \tag{1.27}$$

Equation 1.27 is a generally valid relationship between the noise figure and the effective noise temperature. Note that in the limit of a perfect amplifier or perfect device which adds no noise, the effective noise temperature is 0 and the noise figure is 1.#

The noise properties of mixers are sometimes described in terms of *noise temperature ratio*. This is the ratio of the available noise output power to kT_0B. It may be expressed in dB.

1.3 Noise Temperature and Noise Figure of Attenuators.

In dealing with amplifiers, mixers, or other active components the noise figure is part of the specification for the device. However, noise figures for passive devices such as attenuators need not be specified because they are deducible from basic principles. Attenuators need not be specific devices designed to attenuate microwave signals, but can be any device which attenuates the signal accidentally or otherwise. Thus, for example, wave guide and coax transmission lines are attenuators and as such, contribute some noise which often must be taken account of.

For convenience we will express the attenuation constant of the attenuator as a power ratio and designate that ratio by the symbol α. The attenuation in dB is thus

$$\text{dB} = 10 \log \alpha \qquad (\alpha > 1 \text{ for an attenuator}) \tag{1.28}$$

To derive the noise figure of an attenuator, consider that it has a matched resistance at its input and that the attenuator and the resistance are both at temperature T_1. The configuration is shown in Fig. 1.2.

#Chapter 3 of Ref. 3 may be useful in clarifying the concepts of noise figures and related subjects.

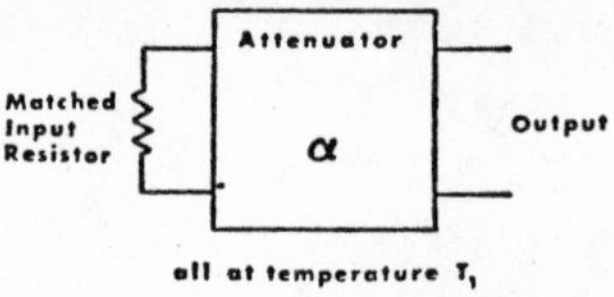

Fig. 1.2. Schematic for calculating noise figure of attenuator.

We can reason from thermodynamics that the available output noise power of the attenuator must be kT_1 per unit of bandwidth, because if it is different from that we could get a net heat exchange from the attenuator output terminals to a load resistor at the same temperature. We can separate this output power in the same way as we did for an amplifier, i.e., one part of it is the amplified or attenuated input noise, and one part is the added noise. Since the input resistor is at temperature T_1, the attenuated input noise is kT_1/α per unit of bandwidth. Therefore, to make up the required output noise, the attenuator must be adding an amount of noise equal to $kT_1(1-1/\alpha)$ per unit of bandwidth. If we refer this back to the input by multiplying by α, we have the quantity which we previously designated as ΔN for an amplifier (in Eq.1.23). In this case

$$\Delta N = (\alpha - 1)kT_1 \text{ per unit of bandwidth} \tag{1.29}$$

Thus; the effective noise temperature of an attenuator from Eq.1.24 is given by:

$$T_e = (\alpha - 1)T_1 \tag{1.30}$$

where T_1 = temperature of tne attenuator.

Using Eq.1.27 we can now calculate the noise figure of the attenuator.

$$F = 1 + (\alpha - 1)\ T_1/T_0 \tag{1.31}$$

Notice that *if the attenuator happens to be at temperature T_0, its noise figure is simply* α. Also notice that in Eq.1.30 if α is very large compared to 1, the effective noise temperature of the attenuator may be very much larger than its actual temperature. In fact, it approaches α times the actual temperature.

Any *real power losses in a passive microwave system can generally be treated as though they were due to an attenuator.* This is true even for absorption losses which occur in a propagation path between antennas or between a microwave source and a receiving antenna. *It is not, however, true of spreading loss* which is really not a loss in this sense of absorbtion. Thus, in radio astronomy for example, a small amount of absorbtion in the atmosphere has a very significant effect upon the performance of the system, not because some of the signal is lost, but because the absorption takes place in a relatively high temperature medium. To illustrate this, let us consider a specific case.

The background sky temperature which a radio astronomy antenna looks at is likely to be of the order of 4°K, while the atmosphere which the signal must traverse may be of the order of 290°K. We can consider this by analogy as a resistor at 4°K at the input to an attenuator whose temperature is 290°K. In the limit of low value of attenuation (0 dB or $\alpha = 1$) the noise added by the attenuator will be 0 regardless of its temperature. The effective temperature of a 0 dB attenuator as given by Eq.1.30 is 0°. On the other hand, if there is 1 dB of attenuation in the atmosphere, $\alpha = 1.26$ and

$$T_e = .26\ T_\circ = 75.5°K \tag{1.32}$$

The noise out of the attenuator (i.e., into the antenna) is then, according to Eq.1.25,

$$\text{Noise out} = k\ B_n \frac{1}{1.26} (4 + 75.5) = k\ B_n\ 63° \tag{1.33}$$

which is about 12 dB more than it would be if there were no atmospheric attenuation present. Thus, in any microwave system which receives noise power from a low temperature source, a small amount of attenuation which takes place in a high temperature medium can have a very great effect upon the system performance.

1.4 Cascades.

One is often interested in finding an overall noise figure for a group of cascaded components, particularly such components as a mixer, an amplifier, and other miscellaneous components which may constitute a receiver. This is mainly a matter of convenience. One can always find the noise output by straightforward calculation, simply considering that each component amplifies or attenuates the noise from the preceeding component and adds some noise of its own. In fact, that is how the cascade noise figure is derived. Consider the series of components shown in Fig.1.3. The gains may be greater than, or less than, 1.

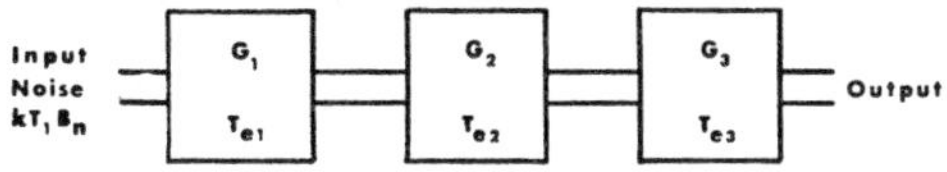

Fig. 1.3. Cascade.

Assume for the moment that the noise bandwidths of all the components are the same and are centered at the same frequency. From Eq.1.25 the noise out of the first amplifier is giver by:

$$k\ B_n\ G_1\ (T_1 + T_{e1}).$$

Using this as input to the second amplifier, the noise out of the second amplifier is given by:

$$k\ B_n\ G_1\ G_2\ (T_1 + T_{e1}) + k\ B_n\ G_2\ T_{e2}.$$

The noise out of the third amplifier is given by:

$$\text{Noise out} = k\,B_n\,[G_1\,G_2\,G_3\,(T_1 + T_{e1}) + G_2\,G_3\,T_{e2} + G_3\,T_{e3}] \qquad (1.34)$$

To find the noise temperature of the cascade, we consider the three components as a single component with noise output given by Eq.1.34, input noise temperature T_1 and gain $G_1\,G_2\,G_3$. Straightforward comparison with Eq.1.25 gives

$$T_e = T_{e1} + \frac{T_{e2}}{G_1} + \frac{T_{e3}}{G_1 G_2} + \ldots\ldots. \qquad (1.35)$$ m

The extension to a larger number of components is obvious.

If we solve Eq.1.27 for T_e and substitute it in Eq.1.35 both for the effective temperatures of the individual components and for the effective temperature of the overall receiver, we arrive at Eq.1.36 which relates the noise figure of the cascade of components to the noise figures of the individual components.

$$F = F_1 + \frac{F_2 - 1}{G_1} + \frac{F_3 - 1}{G_1 G_2} + \frac{F_4 - 1}{G_1 G_2 G_3} + \ldots. \qquad (1.36)$$ m

In the treatment to this point we have assumed that the noise bandwidths of the individual components are the same and that they pass the same frequencies, i.e., that they are centered at the same point and have the same shape. This, of course, is rarely the case. However, it is usual that the last component, generally an amplifier, has the narrowest bandwidth of the group; in which case all the noise outside its bandwidth is of no consequence since it will not pass through to the final output. The entire system can then be treated as though it had the bandwidth of this final amplifier. *cases in which the last circuit component is not the narrowest usually require special treatment,* because the noise will not be uniform or white over the band pass of the final component. In these cases the frequency band pass of the last component must be divided into sub-bands and each one treated separately. The total noise power at the output is, of course, the summation over these individual bands.

CHAPTER II

STATISTICS AND ITS APPLICATIONS

2.1 Statistical Concepts, Binomial Distribution.

Thus far we have been concerned only with thermal noise and only with average values like r.m.s. voltage and current, and the average available noise power. We have characterized thermal noise as being "white", meaning that the average power per unit of bandwidth is independent of frequency. However, we have said nothing about the statistics of this noise, i.e., the extent of the variations above and below the mean value.

It was stated at the beginning of Chapter I that thermal noise is the basic limitation of microwave systems, and instinctively we know that when the signal becomes comparable to or smaller than the noise in its bandwidth, we cannot detect it. However, if the noise is sufficiently steady we should somehow be able to subtract it out before we look for the signal and thus detect a signal which is very much smaller than the noise. After all, the signal adds to the output, and that addition would not be there if the signal were not there. Viewed from this point of view it is not the noise itself which constitutes the basic limitation, but rather its unpredictable statistical variation.

Generally, when a circuit achieves some signal processing gain, what it actually does is somehow increase the ratio of signal effect to the effect of the unpredictable variation in noise level so that we can detect the variation in output due to a smaller signal. To understand this subject of signal processing and processing gains, we must start with the statistics of noise.

Consider a very understandable type of detection problem. This is the problem of detecting whether or not a pair of dice is loaded by noticing the number of 7's which it produces when thrown. Note first of all that the probability of a 7 turning up when a pair of unloaded dice are thrown is 1/6. This can be readily seen by noting that there are 6 numbers which can come up on each of the dice, making a total of 36 combinations, 6 of which add up to 7. Thus the probability of throwing a 7 is 6/36 or 1/6.

Suppose a pair of dice are so loaded that they will produce on the average two 7's out of six throws instead of one. We could detect this if we have a sufficient number of throws of the dice. The question, however, is; what constitutes a sufficient number, and of course that depends in turn upon the amount of confidence we want to have (in a statistical sense) that the dice are loaded or honest before we so state.

Mathematically, the key to the problem is to find an expression which gives the probability of having 1, 2, 3,...n 7's appearing in n throws of the (honest or loaded) dice. This expression is

$$P^n_{(x)} = C^n_x \, p^x \, q^{n-x} \qquad (2.1)$$

This is the *binomial (or Bernouli) distribution#* where

#For further details see, for example, Ref. 4 or 5.

$P^n_{(x)}$ = probability of exactly x 7's in n throws

C^n_x = binomial coefficient $\frac{n!}{x!\,(n-x)!}$ where n! designates factorial n

p = probability of a 7 in a single throw (1/6 or 1/3)

q = probability of anything but a 7 in a single throw (5/6 or 2/3).

With the aid of Eq.2.1 we can answer questions of the type, "Suppose I have 6 throws of an honest pair of dice, what is the probability of 0, 1, 2, 3, 4, 5, and 6 sevens?" And we could ask the same question about the loaded dice. In each case we would get some kind of a distribution with a mean value and a spread. Hopefully we could look at the two distributions and find some number of sevens such that 6 throws of the honest dice usually produced less than that number and the loaded dice usually more. Then we might use that number as a decision criterion.

To increase the sensitivity of our test we could simply use more throws. If we used 60 instead of 6, we would expect about 10 sevens from the honest dice and 20 from the loaded dice; and although the spreads of each distribution would be numerically larger than for 6 throws, we would expect it to be a smaller fraction of the difference between the mean values (20 minus 10). Perhaps 15 sevens in 60 throws would be a good criterion.

The problem we are discussing can be related directly, element for element, to the signal detection problem by considering that the presence of a signal is analogous to the average number of 7's per throw being different from its nominal value (1/6). The statistical variation is inherent in the noise. At some point we must make a decision as to whether or not a signal is present. Because of the statistical nature of the situation, *there is always some probability of a false alarm,* i.e., whatever criterion we apply to decide that a signal is present, will occasionally indicate that a signal is present when one is not present just due to statistical fluctuations; and *there is always some probability of a missed detection,* i.e., deciding that a signal is absent when one is really present, due to a downward fluctuation in the noise.

Just as, in the case of detecting loaded dice we can improve our probability of correct decisions by observing a greater number of throws before we decide; in the signal case we can improve our probability of detection and/or decrease our probability of false alarm by observing the output for a longer period of time before the decision is made. In the detection case, however, there is usually some limit to observation time placed upon us by the fact that the signal does not persist forever, but only for some brief period of time. This time may be part of the design of the system. Clearly, some intelligence must be transmitted by having the signal there or not there. In a digital system one might say that a 1 is transmitted when the signal is there, and a 0 is transmitted when the signal is not there. We have to keep the signal there or have it missing for some predetermined period of time, and it is advantageous in one way to make that period of time as short as possible so that the system can transmit a maximum number of bits of information per second. Clearly we have some sort of tradeoff between information rate and the signal to noise ratio at which we can operate.

There is, however, one correspondence between the two detection problems which is fairly subtle, and it is in the area in which most statistical paradoxes have their fallacy. Without saying anything about it, we understand that each throw of the dice is a statistically independent event. By this we mean that the probabilities on any particular throw are quite independent of what number was thrown last. The corresponding property in signal detection would be some length of time over which there is statistical independence of the same sort. If the voltage happens to be high at one instant in time, we understand that surely there is some brief period of time over which it is likely to remain high. Thus, to take an extreme example, if I have a signal with frequency content up to about 1 kHz, which happens to be high at some instant of time, I can be fairly confident that 1 microsecond (μsec) later it will still be high. Thus, two samples of that signal taken 1 μsec apart are clearly not statistically independent. It can be shown that the *time interval required between two samples of noise in order that they be statistically independent is of the order of* $1/2B_n$ where B_n is the noise bandwidth from Chapter 1. m

One way to convince yourself of this qualitatively is to consider that the noise is white and sharply cut off at some upper and lower frequencies. Let it then be heterodyned to a lower frequency so that the lower edge of the bandwidth is at zero frequency. Its highest frequency is then B_n and by Shannon's theorum# it can contain at most $2B_n$ independent samples per second.

Returning to our binomial distribution problem; Eq.2.1 is a precise means of calculating anything we want to know about the probable number of 7's or any similar distribution although it is quite cumbersome. We would like answers to the following questions. If I throw a pair of unbiased dice n times, what is the average number of 7's which I expect, and what standard deviation or spread do I expect about that mean value? Notice that when I talk about the average and the standard deviation I am somehow implying that I am going to do this experiment many many times; the experiment being n throws of the dice.

The binomial distribution is so called because the probabilities are given by the binomial theorum. Thus

$$(p+q)^n = \sum_{x=0}^{x=n} C_x^n\, p^x\, q^{n-x} = \sum_{x=0}^{x=n} P^n_{(x)} \tag{2.2}$$

The first equality in Eq.2.2 is simply the binomial theorum. The second equality comes from Eq.2.1. Since $p+q=1$, we are assured by virtue of Eq.2.2 that the sum of all the probabilities is 1.

This relationship is used in appendix D to show that the mean (or expected) value of x is given by

$$\overline{x} = np \tag{2.3}$$ m

and the variance is given by

$$\sigma^2 = npq \tag{2.4}$$ m

Notice that when p is small, i.e., when the probability of some particular outcome of interest in a single event is small, and q is therefore close to 1, Eq.2.4 indicates that the variance (σ^2) is equal to the mean.

Thus, for example, consider nuclear decays as counted by a Geiger counter or scintillation device where each event, which corresponds to a nuclear decay with certain properties, is quite independent of any other event and has very small probability of occurrence in a time interval of interest. Suppose one finds a mean number of counts occurring in some interval of time, then the *variance in the number of counts observed in that interval of time will be equal to that mean* and the standard deviation will equal the square root of that variance. Thus, for example, if one has 100 counts per minute one would expect a variance of 100 counts per minute and hence a standard deviation equal to the square root of that, or 10 counts per minute. Thus

---

#Shannon's sampling theorum (Ref.6) states that if a function has frequency content only between 0 and f cycles per second, it is completely determined by giving its ordinates at points spaced 1/2f seconds apart. The situation is analogous to fitting points with a polynomial curve.

one would expect the count rate to be between 90 and 110 about 2/3 of the time.#

One of the important characteristics of statistical distributions is their property under addition and subtraction. If we add randomly selected samples of two statistically independent distributions to form a new distribution, the distribution of the sums will have a mean value equal to the sum of the means and a variance equal to the sum of the variances (Ref. 7). If we subtract instead of adding, the mean will be the difference of the means, but the variance will still be the sum of the variances. Thus given that

$$\sum_i (a_i - \bar{a})(b_i - \bar{b}) = 0 \ . \tag{2.5}$$

If

$$c_i = a_i + b_i \ , \text{ then } \bar{c} = \bar{a} + \bar{b}$$

and

$$\sum_i (c_i - \bar{c})^2 = \sum_i (a_i - \bar{a})^2 + \sum_i (b_i - \bar{b})^2 \tag{2.6}$$

If

$$c_i = a_i - b_i \ , \text{ then } \bar{c} = \bar{a} - \bar{b}$$

and

$$\sum_i (c_i - \bar{c})^2 = \sum_i (a_i - \bar{a})^2 + \sum_i (b_i - \bar{b})^2 \tag{2.7}$$

where eq.2.5 is the mathematical consequence of the fact that the samples are selected randomly.

Before leaving the problem of detecting whether or not a pair of dice are loaded, let us consider it in a more general way. Assume that we have a fairly large number (n) of throws of the dice on which to base our judgement. Clearly any number of 7's from 0 to n are possible with either loaded or unloaded dice, but the probability distribution which we expect is quite different. Figure 2.1 shows schematically the types of distributions which we might expect from loaded and honest dice.

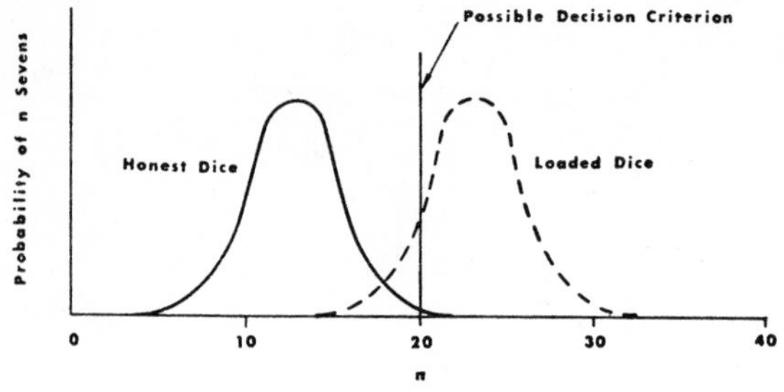

Fig. 2.1. Illustration of decision criterion.

#The probability of falling outside $\pm$ 1σ from the mean is .32 in a normal or Gaussian distribution and it will be about the same here by virtue of the central limit theorum to be discussed in Section 2.3.

One of the things that makes probability concepts a little difficult to grasp is the implication of many trials. Thus if we state the probability P of 15 sevens in 60 throws of the dice, the number tells us what will happen if we do the experiment many times; the experiment being to throw the dice 60 times.

$$P = \lim_{N\to\infty} \frac{N_{15}}{N}$$

where N_{15} = number of times we get 15 sevens

N = number of times we try 60 throws.

We have to adopt some decision criterion based on one experiment (not one throw of the dice). The signal detection problem is the same in this respect. Any such criterion, for example the one shown in Fig.2.1, will have some probability of false alarm (this is the area under the solid curve to the right of the decision criterion) and some possibility of missed detections (this is the area under the dotted curve to the left of the decision criterion).

In many detection problems there is room for argument about what constitutes an optimum detection criterion. It depends on the consequences of detection and nondetection as well as the *apriori* probability that the signal is present. This is the fraction of the cases in which we expect it to be there. In the case of the dice we would take account of this by some instinctive feeling, e.g., are we playing with friends or strangers? Do they look honest? And what will happen if we accuse someone of cheating? In most communication work the consequences of a missed detection are, in significance, equal to those of a false alarm since the consequence of mistaking a 0 for a 1 is usually the same as mistaking a 1 for a 0. Further, ones and zeros are equally likely. If one makes these assumptions the optimum detection criterion is then that which minimizes the sum of the two error areas. Generally such a criterion also equalizes the two areas, i.e., the probability of a missed detection will be equal to the probability of a false alarm. In radar, for example these assumptions are grossly invalid. We will have more to say on this subject in Section 2.6.

A word about statistical nomenclature is in order. A relationship like Eq.2.1 which gives the probability of occurrence of particular values of the variable x is called a *mass function*. If it were a similar function for a continuous distribution (e.g., the normal distribution) it would be called a *density function*. The integral from zero to x of a mass function or a density function is called a *distribution function* or probability distribution function. It gives the probability that the value is less than (or equal to) x. Also the variable x is usually called a *random variable* although there are some subtleties about that last term that I prefer to avoid discussing.

The hypothesis that no signal is present is usually identified as the *null hypothesis*. In that case a false alarm is what statisticians call a *type I error* and a missed detection is a *type II error*. The mean value of a random variable is the same as its *expected value*, written E(x) and the *variance*, is usually denoted as V(x) or σ^2 where σ is the *standard deviation*.

2.2 Shot Noise

Shot noise will be discussed at this point because it is a purely statistical phenomenon which is readily derivable from the binomial distribution and more particularly from the limiting case of that distribution where n is very large and p is very small. Under these circumstances the distribution becomes *Poisson*.

The important characteristics of any shot noise generator are that electrical charge is transported in a finite sized package and that *the packages travel independently of each other*. The easiest case to visualize is perhaps the temperature limited vacuum tube diode. In this case electrons leave the cathode independently whenever they reach the surface of the cathode with sufficient kinetic energy to overcome the barrier potential. When they leave they are picked up by the electric field and transported across the diode.

The fact that they do not interact with each other is expressed by saying that the diode is temperature limited rather than space charge limited. This implies that the current across the diode is limited by the temperature of the cathode, i.e., that there is ample voltage across the diode to sweep up all the electrons which come out of the cathode and transport them to the anode and that the only limitation of the current is that the temperature is low, so that it is only the extreme of the thermal distribution which provides any probability of electrons leaving the cathode.

In this situation the probability of any particular electron transiting from the cathode to the anode is independent of the probability of any other electron doing so. Thus we have statistical independence and we can treat each electron in the cathode as having some extremely small probability p of leaving the cathode per unit of time and transiting to the anode. Therefore we can write

$$\text{Average number of electrons transiting in time } t = npt \tag{2.8}$$

where n = number of electrons in the cathode

p = probability of each individual electron leaving the cathode per second.

In principal we could apply Eq.2.1 to calculate the probability of any particular number of electrons transiting in time t, but in practice this is quite impossible since n is of the order of magnitude of Avagadro's number (the number of atoms in a gram molecular weight of material). This is about 10^{22}. However, we can use the derived fact that for such a distribution the variance is equal to the mean (since q, or the probability of any particular electron not being emitted in time t is very close to 1). Thus the standard deviation in the number of electrons making the transit in time t is given by Eq.2.4 as

$$\sigma \text{ in number of electrons} = \sqrt{npt} \tag{2.9}$$

If we multiply Eq.2.8 and 2.9 by the charge of the electron and divide by the time interval, we obtain the average current and the standard deviation of that current. The result is

$$\text{Average DC current } I = npe \tag{2.10}$$

$$\sigma_{current} = e\sqrt{np/t} \tag{2.11}$$

We cannot, of course, measure n or p, but we can solve Eq.2.10 for the product and substitute this in Eq.2.11. Thus,

$$\sigma_{current} = \sqrt{Ie/t} \tag{2.12}$$

This result can probably best be interpreted in terms of the following experiment. I have a temperature limited diode and a meter which measures the current through it. The meter, however, measures current for some time interval t and reports the average value of the current during that time. It then goes on to the next time interval t and reports the average value of the current during that time, etc. Equation 2.12 says that the reported values will have a standard deviation about their mean which is proportional to the square root of the mean value, the square root of the charge of the electron, and inversely proportional to the square root of the time over which the average is taken. σ is, by definition, the r.m.s. deviation from the mean. Therefore in the limit as t becomes small this variation in current becomes the a.c. component of the current and Eq.2.12 gives its r.m.s. value. The available a.c. shot noise power, by analogy with the discussion following Eq. 1.1, is therefore IeR/4t, where R is the internal resistance of the shot noise generator.

An important remaining problem, however, is to determine the frequency content of that a.c. current. That is, over what bandwidth is it spread, and how does it vary with frequency. Notice that the variation is based upon 1/t measurement per second; which provide 1/t independent values per second and therefore, by Shannon's theorum# are capable of defining a signal with frequency content from zero to 1/2t. Therefore it is reasonable to accept the fact that the frequency content of the a.c. current is between f=0 and f=1/2t. This could be proven with rigor.

To investigate the way in which the shot noise energy is spread out over this frequency band, consider a specific set of average current measurements made as described above at intervals of 1/t sec; to which we will fit a curve of current vs time and then find the frequency content from the Fourier or LaPlace transform of that curve. If we average these measurement points together by pairs before we fit the curve, the transform will be the same in the low frequency half of the spectrum, but the high frequency half will disappear. But averaging the measurements together by pairs is the same as doubling the measurement time. As can be seen from the relations just derived, this will cut the available shot noise power in half as well as cutting the bandwidth in half. Therefore half the shot noise power is in the lower half of the frequency band and half in the upper half. Similarly we can show that one quarter of the power is in the lower quarter of the band, etc. This means the noise power is uniformly spread and is white. Therefore the power per unit bandwidth is

$$\text{Available shot noise power per hertz} = \frac{IeR/4t}{1/2t} = eIR/2 \qquad (2.13)$$ m

where e = charge on the electron (1.6×10^{-19} coulombs)

I = average d.c. current

R = internal resistance of the shot noise generator.

Concerning shot noise power magnitudes; if the current is a milliamp, the resistance is 1000 ohms and the bandwidth is 1 megahertz, one can readily show (by Eq.2.13) that the power will be 8×10^{-14} watts, corresponding to a voltage of about 9 microvolts. Compare this with the 4 microvolts of thermal noise which we found across a 1000 ohm resistor at room temperature with the same bandwidth. There is nothing significant about the exact ratio of these two voltages, however, since if the diode were carrying 100 milliamps instead of 1, its shot noise power would be 10 dB greater.

Equation 2.13 provides us with a way of *measuring the charge on the electron.* All we have to do is measure the noise power out of a temperature limited diode in some bandwidth when the diode is carrying some known value of current. Actually, this is a viable method and has been used for that purpose. The problem, however, is one of obtaining adequate accuracy for scientific purposes.

The problem of *eliminating shot noise* is simply the problem of eliminating statistical independence. By operating in space charge limited conditions rather than temperature limited conditions, we destroy the statistical independence of the transit of the charges, and since the charges interact with each other by repulsion, the current is smoothed out.

#See section 2.1.

2.3 The Gaussian Distribution

The Gaussian Distribution is very important in many fields of science, mainly because of the *central limit theorem* which states that, given an adequate degree of randomness, most continuous distributions and envelopes of the mass functions of discontinuous distributions tend to be Gaussian (Ref.8). It is for this reason that it is also called the normal distribution. Thermal noise and shot noise are Gaussian in the following sense. If we take samples of instantaneous thermal noise voltage or shot noise voltage, we will find that they define a probability distribution of Gaussian shape. For example, the probability of the voltage being within some small interval near its mean is more likely than finding it within the same size interval one σ away from its mean by exactly that ratio which is given by a table of normal distributions (.607). Similarly, the probability of finding it 3 times the r.m.s. voltage (3σ) from its mean is .011 times the probability of finding it near its mean, etc.

The Gaussian or normal distribution is defined by

$$p(x)\ dx = \frac{1}{\sqrt{2\pi}\ \sigma} \exp - [(x-\bar{x})^2/2\sigma^2]\ dx \qquad (2.14)$$

It is shown in appendix E that the mean and standard deviation of the distribution are in fact $\bar{x}$ and σ as indicated by Eq.2.14.

The *DeMoivre-LaPlace limit theorum* relates the Gaussian distribution to the binomial distribution. It states that under suitable conditions

$$P_B\ \{\alpha < x < \beta\} \approx \frac{1}{\sqrt{2\pi}\ \sigma} \int_{\alpha-1/2}^{\beta+1/2} \exp - \frac{(y-\bar{y})^2}{2\sigma^2}\ dy \qquad (2.15)$$

where $P_B\ \{\alpha < x < \beta\}$ = Binomial probability calculated from Eq.2.2 that x, which can take only integer values, is between α and β (both integers)

$\sigma = \sqrt{npq}$ as in Eq.2.4

$\bar{y} = np$ as in Eq.2.3.

Equation 2.15 is valid if $\frac{\alpha^3}{\sigma^4}$ and $\frac{\beta^3}{\sigma^4}$ are small (Ref.5). It says that *the integral of the binomial mass function and the integral of the Gaussian density function between essentially the same limits is about the same if both distributions have the same means and standard deviations.*

2.4 Statistics of Detector Outputs

The decision as to whether or not a signal is present is never made by directly examining an r.f. signal, but rather by detecting (or rectifying) it first. It is therefore important to understand the statistics of rectified or detected noise. In general, the original r.f. noise will be confined to some frequency band which does not extend to dc. This is what is meant by narrow band noise. We can express such narrow band noise as a modulated signal with the statistical, noiselike character confined to the modulation. Thus

$$n(t) = n_1\ (t) \sin \omega_0 t + n_2(t) \cos \omega_0 t$$

$$= r(t) \exp-(j\omega t)$$

$$r(t) = \sqrt{n_1^2(t) + n_2^2(t)} \tag{2.16}$$

If n(t) is Gaussian, symmetric about ω_0 and has variance σ^2, then $n_1(t)$ and $n_2(t)$ are Gaussian and statistically independent of each other and each has a variance σ^2 (Ref.9).

It is not surprising that the two modulations are Gaussian, but the other two facts given about them are surprising. First, consider the fact that they both have variance σ^2. Since the original r.f. signal had variance σ^2, this would seem to mean that the power in the original signal has somehow been doubled. However, one can readily show that the power in the modulated signal is only half of what it would be if the wave form were a signal itself instead of a modulation. Thus the total power is preserved. The fact that the two modulations are statistically independent is even more surprising since they are both derived from the same noise signal. However, this is a consequence (although not an obvious one) of the fact that the carrier is taken at the center of a symmetrical band of noise. In most practical cases the modulation signal is derived by a detector following the i.f. amplifier and the assumption of symmetry amounts to an assumption that the noise is uniform across the bandwidth of the i.f. amplifier. If this is not true, then the results to be given are not rigorously correct and some small deviations may be expected. Assuming proper symmetry, however, we can write for the probability of any combination of values of n_1 and n_2,

$$P(n_1,n_2)\,dn_1\,dn_2 = P(n_1)\,p(n_2)\,dn_1\,dn_2$$

$$= \frac{1}{2\pi\sigma^2} \exp - \left[\frac{n_1^2}{2\sigma^2} + \frac{n_2^2}{2\sigma^2} \right] dn_1\,dn_2 \tag{2.17}$$

but $n_1^2 + n_2^2 = r^2$

and as in the derivation given in appendix E we can replace $dn_1\,dn_2$ by $r\,dr\,d\theta$, integrate over θ and find

$$p(r)\,dr = \frac{r}{\sigma^2} \exp - \left[\frac{r^2}{2\sigma^2}\right] dr \tag{2.18}$$

This is a *Rayleigh* density function. It is the probability distribution of the voltage from a linear or envelope detector when the input is narrow band noise. It has the general shape shown in Figure 2.2.

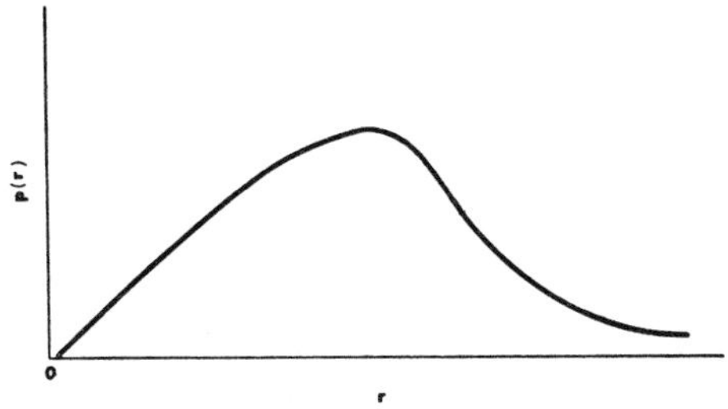

Fig. 2.2. Rayleigh density function

We can readily find the mean value of this output voltage and its a.c. component (i.e., the mean square deviation from the mean). The mean value is given by

$$\overline{r} = \int_0^\infty \frac{r^2}{\sigma^2} \exp - [r^2/2\sigma^2]\, dr \tag{2.19}$$

Integration by parts gives

$$\overline{r} = \sigma \sqrt{\pi/2} \tag{2.20}$$

The a.c. component is given by (using Equation D.4 of appendix D)

$$\sigma^2{}_r = \int_0^\infty \frac{r^3}{\sigma^2} \exp - [r^2/2\sigma^2]\, dr - \frac{\pi}{2}\sigma^2$$

and integration by parts gives

$$\sigma_r^2 = (2 - \pi/2)\, \sigma^2 \tag{2.21}$$

Combining Eq.2.20 and 2.21

$$\frac{\sigma_r}{\overline{r}} = \frac{\sqrt{2 - \pi/2}}{\sqrt{\pi/2}} \cong .525 \tag{2.22}$$

Note that this is quite different from what we found for the binomial distribution. There the variance was equal to the mean. In this case it is the *standard deviation that is proportional to the mean* and the variance is proportional to the square of the mean.

The distinction is quite significant and, in fact, it is the only one which could correspond to reality. If the standard deviation were proportional to the square root of the mean in the noise output of a detector, then one could cut down the precentage fluctuation in the output by simply working with a larger signal, that is, by having a few more stages of gain in the i.f.. amplifier. By cutting down this standard deviation, one would be able to detect a smaller signal superimposed on the noise. Thus, one could achieve any desired level of signal processing by simply using more i.f. gain. This, of course, is not the way the world works.

We will compare Eq.2.22 with similar results for other detectors. First consider the available power out of such a linear detector.

$$\text{power} = W = r^2/4\rho$$

where ρ = is the video output resistance of the detector. It follows that $dW = r\,dr/2\rho$.

If we use these two equations to eliminate r and dr in Eq.2.18, and to acheive more conventional notation, we let $R=4\rho$, we find the density function

$$p(W)\,dW = \frac{R}{2\sigma^2}\ \exp - \left[\frac{WR}{2\sigma^2}\right] dW \qquad (2.23)$$ m

This is the *Rayleigh power density function* which looks like Figure 2.3.

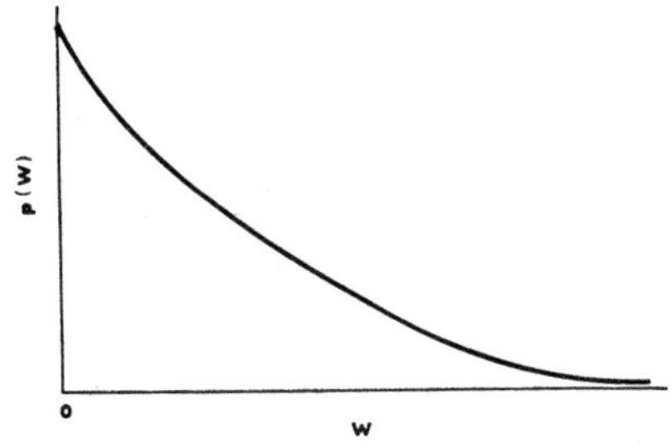

Fig. 2.3. Rayleigh power density function.

The mean value is given by

$$\overline{W} = \int_0^\infty W\,p(W) = \frac{R}{2\sigma^2}\int_0^\infty W \exp - [WR/2\sigma^2]\,dW$$

Integrating by parts, letting $W = u$ and $\exp - [WR/2\sigma^2] = dv$. we find

$$\overline{W} = 2\sigma^2/R \qquad (2.24)$$

To find the mean square deviation from the mean

$$\sigma_W = \frac{R}{2\sigma^2}\int_0^\infty W^2 \exp - [WR/2\sigma^2]\,dW - 4\sigma^4/R^2 \qquad (2.25)$$

where we have used the integral analog of Eq.D4 of appendix D.

Again integrating by parts, we find

$$\sigma_W^{\ 2} = 4\sigma^4/R^2 \qquad (2.26)$$

From Eq.2.24 and 2.26

$$\frac{\sigma W}{\overline{W}} = 1 \qquad (2.27)$$

Compare this with Eq.2.22.

where W = instantaneous output power of the detector

$\overline{W}$ = average output power

σ^2 = input r.f. noise power (with proper normalization)

R = 4 times the video output resistance of the detector.

Next let us examine the square law detector. We can write the square law detector voltage output as

$$V_S = \overline{[n(t)]^2} = \overline{(n_1(t) \sin w_0 t + n_2(t) \cos w_0 t)^2}$$

$$= \frac{1}{2} n_1{}^2(t) + \frac{1}{2} n_2{}^2(t) \qquad (2.28)$$

But this is just 1/2 $[r(t)]^2$ where r(t) is given by Equation 2.18. Therefore, the voltage out of the square law detector will have exactly the same distribution as the power out of the linear detector and consequently, as shown in Eq.2.27, the mean voltage out of the square law detector will be equal to the standard deviation of that voltage.

Next, for illustrative use in the next section rather than for practical value, let us look at the power out of a square law detector. Since this is proportional to the square of the voltage out of such a detector, it is proportional to the square of the power out of a linear detector, and one can write from Eq.2.23

$$W_S = W^2, \; dW_S = 2WdW$$

$$P(W_S) = \frac{R}{4\sigma^2\sqrt{W_S}} \exp - \left[\frac{R\sqrt{W_S}}{2\sigma^2} \right] dW_S \qquad (2.29)$$

Again we can find the statistical properties in a straightforward manner. (To integrate it is convenient to substitute $Z^2 = W_S$.) The results are

$$\overline{W_S} = 8\sigma^4/R^2 \qquad (2.30)$$

$$\sigma_{W_S} = 8\sqrt{11}\,\sigma^4/R^2 \qquad (2.31)$$

and

$$\frac{\sigma_{W_S}}{\overline{W_S}} = \sqrt{11} \doteq 3.32. \qquad (2.32)$$

Notice that whether we use a linear or square law detector, and whether we consider the output voltage or output power, the standard deviation is a fixed multiple of the mean, rather than being proportional to the square or square root of the mean. The ratio of standard deviation to mean in the cases considered ranged from 0.525 to 3.32. Offhand it would seem desirable to choose that particular type of detector and output to work with which has the smallest ratio of standard deviation to mean since this will provide the most constant output against which to detect the increment due to a signal. However, the addition of a signal of the same magnitude will not change all the detector outputs by the same amount. This subject is covered in the next section.

2.5 Choice of Detector and Output Type

This is a subject about which one can very easily become confused. Most treatments of the subject which appear in the literature are too mathematically elegant to provide the student with a valuable feel for the problem. There are two factors which determine the merit of a particular type of detector. These are the statistics of its output as calculated in Section 2.4 which can be characterized by a sigma or rms fluctuation, and the change in average output which occurs due to the signal. The larger the ratio of change to fluctuation for a particular signal to noise ratio, the better the detector. We will define the figure of merit of a detector as the dimensionless quantity

$$\text{FOM} = \frac{\Delta}{\sigma_0 \eta} \tag{2.33}$$

where Δ = change in average output when a signal of power content S is added to an average noise power of N

σ_0 = rms fluctuation in output#

η = signal to noise power ratio at the detector input (S/N)

We may clearly expect that the type of detector whose output changes most with the addition of a small signal may not be the one which changes most when a large signal is added. Δ is generally not simply proportional to the signal power. Thus the detector type which has the best figure of merit for small signal to noise ratios may not be best for large signal to noise ratios.

In this section we will calculate the figure of merit, defined by Eq. 2.33, of 2 common types of detectors; linear and square law. For illustrative purposes rather than for practical value we will also calculate the FOM of a square law detector whose output is squared before being presented to decision circuits. Detailed calculations are presented only for the limiting case of small signals but more general results are given. For each detector type we will calculate Δ and σ_0 in terms of the average detector output, which will cancel out in the final formula for FOM.

In Section 2.4 we calculated three ratios of detector output fluctuations to mean output (with noise alone present as input) ($\sigma_0/\overline{W}$ or $\sigma_0/\overline{r}$). These were .525 for the voltage output of a linear detector, 1.0 for the power output of a linear detector or the voltage out of a square law detector, and 3.32 for the power out of a square law detector. We will now calculate, for a small signal, the ratio of change in output to average output for each of these detectors.

The only case which presents any mathematical complexity is the linear detector. Thus far we have examined the detector output only when no signal is present. When signal is present, the detector output density function is given by (Ref.10).

#For the same noise input the rms fluctuation may change when the signal is added. We should properly take account of this, but for simplicity we will consider only the rms fluctuation with the signal absent.

$$P(r)\,dr = \frac{r}{\sigma^2} \exp - \left(\frac{r^2 + 2S}{2\sigma^2} \right) I_0 \left(\frac{\sqrt{2\,S}\,r}{\sigma^2} \right) dr \qquad (2.34)$$

where r = output voltage (as before)

σ^2 = input noise power

S = r.m.s. signal power

I_0 = modified Bessel function of the first kind of zero order.

For $S=0$, $I_0 \to 1$ and Eq.2.34 becomes Eq.2.18. To find $\overline{r}$ for small S, we multiply Eq.2.34 by r and integrate from zero to ∞. The integration is difficult but results have been tabulated (Ref.11). For small signal to noise ratios the increase in detector output when the signal is added is given by

$$\Delta_0 = \overline{r}\,\eta/2$$

where η = signal to noise power ratio.

Thus our figure of merit for this type of detector for small signals using Eq.2.22 is

$$\text{FOM (linear, small } \eta) = \frac{\overline{r}\,\eta/2}{.525\,\eta\,\overline{r}} = .95 \qquad (2.35)$$

For the square law detector, the fractional increase in voltage output is simply the ratio of the signal power to the noise power or

$$\Delta_0 = \overline{W}\eta$$

and using Eq.2.27 we find

$$\text{FOM (Square law voltage, small } \eta) = \frac{\overline{W}\eta}{\overline{W}\eta} = 1 \qquad (2.36)$$

The power output of the square law detector is the square of the voltage output which is $(\sigma^2 + S)^2 = \sigma^4 + 2S\,\sigma^2 + S^2$. For small S, the mean output is σ^4 and the term in S is larger than that in S^2. Therefore,

$$\Delta_0 = \overline{W}_s \frac{2S}{\sigma^2} = 2\overline{W}_s\eta$$

and

$$\text{FOM (square law power, small } \eta) = \frac{2\overline{W}_s\,\eta}{3.32\,\overline{W}_s\,\eta} = .6 \qquad (2.37)$$

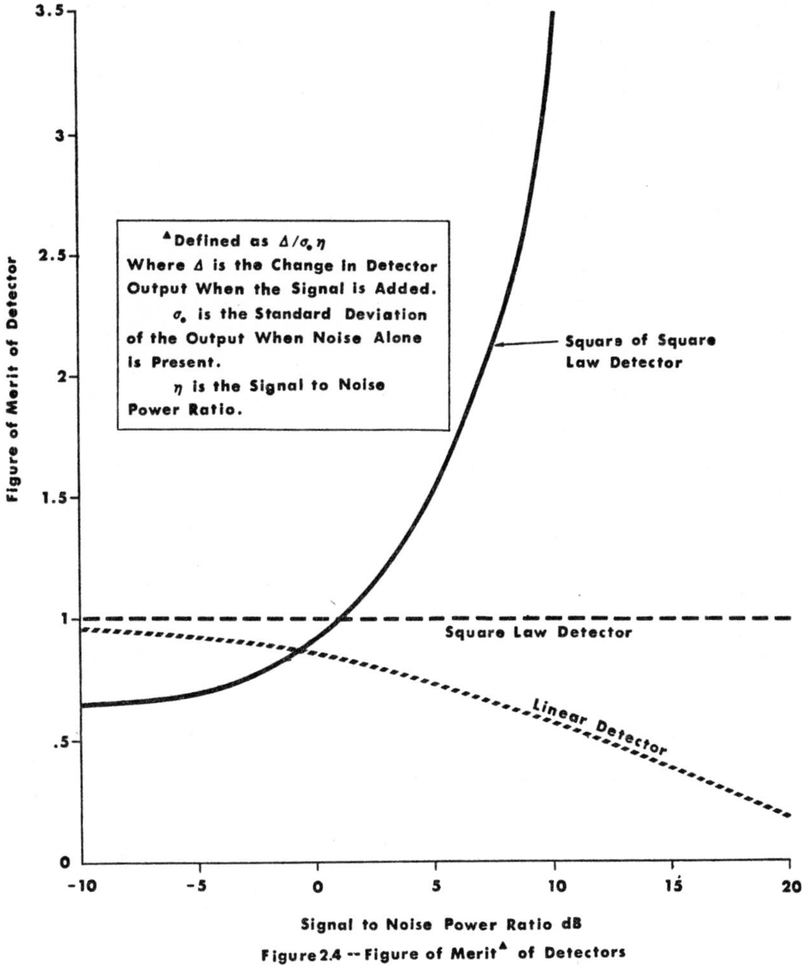

Figure 2.4 -- Figure of Merit▲ of Detectors

Thus for very small signal to noise ratios (less than 1) the square law detector voltage output provides the best Figure of Merit, defined as the ratio of mean output change for a given signal; to random fluctuations. However, it is only 5% (0.2 dB) better than the linear detector.

Figure 2.4 shows detector figure of merit data as a function of signal to noise ratio calculated in this way for the three different detector schemes we have analyzed. From the figure, it would seem that the square law detector is slightly better at low signal to noise ratios, and gains relative to the linear detector as the signal increases. The square law power detector appears best at signal to noise ratios greater than about +1 dB. The data in Fig.2.4 must be further qualified for two reasons. Probably most significant is that the detector outputs are not normally distributed so that the standard deviation in the output may not be an accurate indication of what the low probability extremes of the detector outputs are. In some detection systems it is these low probability extremes which cause false alarms and missed detections. The second point is that the output fluctuation generally increases when signal is present. The increase is not the same for all detection schemes. However, neither of these qualifications is significant when the system is designed to operate with small signal to noise ratios at the detector which is the case where large incoherant (post detection) processing gains are acheived. This is discussed in Section 3.2.

2.6 The Bayes Criterion and Likelihood Fuctions.

One of the fundamental problems of detection theory can be put in the following terms: I have some sort of a signal processor which gives me an output voltage, on the basis of which I must decide whether or not a signal is present. What criterion shall I use? The Bayes Criterion is surely *the optimum basis* for making such a decision provided all the necessary input data can be obtained. It is essentially a minimum cost criterion. We attach a cost to making each of the two possible types of errors, and we also know or assume the apriori probability of a signal being present. If in addition we know the processor output probability distribution functions with and without the signal present#, we can determine the optimum threshold. Thus

$$\bar{C} = (1-p)\, C_{10} \int_{x_T}^{\infty} f_0\,(x)\, dx + p\, C_{01} \int_{o}^{x_T} f_1\,(x)\, dx \qquad (2.38)$$

where $\bar{C}$ = average cost of mistakes, which we wish to minimize

C_{10} = average cost of a false alarm

C_{01} = average cost of a missed detection

p = probability (apriori) that a signal is present

x_T = threshold value of x selected as the basis for deciding whether or not a signal is present

$f_0(x)$ = processor output density function with no signal present

$f_1(x)$ = processor output density function with the signal present.

Since $\int_{x_T}^{\infty} f_0(x)\, dx = 1 - \int_{o}^{x_T} f_0(x)\, dx$, we can write Eq.2.38 as

$$\bar{C} = (1-p)\, C_{10} + \int_{o}^{x_T} [p\, C_{01}\, f_1(x) - (1-p)\, C_{10}\, f_0(x)]\, dx \qquad (2.39)$$

#Analogous to those shown in Figure 2.1.

For sufficiently small values of $x, f_0(x) >> f_1(x)$ and the integrand will be negative. For large values it will be positive. Therefore, the minimum value of $\bar{C}$ (the point at which the integral has its greatest negative value) will occur when the integrand is zero or

$$p\, C_{01}\, f_1(x_T) = (1-p)\, C_{10}\, f_0(x_T)$$

$$\frac{f_1(x_T)}{f_0(x_T)} = \frac{(1-p)\, C_{10}}{p\, C_{01}} \qquad (2.40)$$

The quantity $f_1(x)/f_0(x)$ is called the *likelihood ratio. It is simply the ratio of the ordinates of the two curves* (eg., see Fig.2.1) *at a particular value of x*. Equation 2.40 tells us that the optimum decision threshold is at that point where the ordinates of the two curves have a certain ratio. However, of course, the quantities on the right hand side of Eq.2.40 may be difficult to estimate. Note that if $p=1-p$ and $C_{01}=C_{10}$, which is the usual case in digital communication systems, the optimum ratio of the ordinates is 1, i.e., we should operate at the point where the density functions cross.

2.7 Reliability Theory.

In electronic or mechanical components we are often concerned with *time to failure* (TTF) or *mean time to failure* (MTTF) of if the component is a switch or other item which operates in cycles, we may be concerned with *cycles to failure* (CTF) or *mean cycles to failure* (MCTF). In the former case a graph of the probability of failure as a function of time is a density function (continuous), and in the latter case it is a mass function (integer values only). We will concern ourselves first with the latter case.#

The most common assumption in reliability theory is that the probability of the component failing on any particular cycle or trial is independent of the number of previous cycles on which it has operated correctly. This implies that the component has no memory, no wear, and all trials are independent. The probability of failure on any trial n, or the mass function, is then given by

$$f_C\,(n,q) = p^{n-1}q, \quad n=1,\ 2,\ 3.....$$

where q = probability of failure per trial (independent of n),

$p = 1-q$.

The mean or expected number of cycles to failure is then given by

$$E(C) = \sum_{n=1}^{\infty} n\, p^{n-1}\, q = q \sum_{n=1}^{\infty} n\, p^{n-1} = \frac{q}{(1-p)^2} = \frac{1}{q} \qquad (2.41)$$

where we have recognized that

$$\sum_{n=1}^{\infty} n\, p^{n-1} = 1 + 2p + 3p^2 + 4p^3 + ... = \frac{1}{(1-p)^2} \qquad (2.42)$$

Thus, MCTF = $1/q$

#A good discussion of this subject is contained in Chapter 10 of Ref. 12.

The variance in CTF is given by#

$$V(C) = \sum_{n=1}^{\infty} n^2 p^{n-1} q - (\tfrac{1}{q})^2$$

$$= q [\sum_{n=1}^{\infty} n^2 p^{n-1} - (\tfrac{1}{1-p})^3] = q [\tfrac{1}{(1-p)^3} + \tfrac{p}{(1-p)^3} - \tfrac{1}{(1-p)^3}]$$

$$V(C) = \frac{p}{q^2} \qquad (2.43) \quad m$$

where we have similarly recogiized that

$$\sum_{n=1}^{\infty} n^2 p^{n-1} = \frac{1}{(1-p)^3} + \frac{p}{(1-p)^3}$$

In the continuous case, involving TTF, it is common to make the same assumption of statistical independence and lack of memory. In this case we assume that the probability of failure in any interval of time Δt is independent of time. The resulting density function is the *exponential distribution*

$$f_T (t,\theta) = \theta e^{-\theta t} \qquad (2.44) \quad m$$

Notice that the distribution function is

$$F_T (t,\theta) = \int_0^{t_o} \theta e^{-\theta t} dt = -e^{-\theta t}]_0^{t_o} = 1 - e^{-\theta t_o}$$

and that

$$\int_0^{\infty} F_T (t,\theta) = 1$$

The reliability at time t_o which is defined as the fraction of the components surviving to time t_o is given by

$$R(t_o) = e^{-\theta t_o} \qquad (2.45) \quad m$$

and

$$MTTF = \int_0^{\infty} t \theta e^{-\theta t} = 1/\theta \qquad (2.46) \quad m$$

#See again Eq.D.4 of appendix D.

where we have integrated by parts.

Where components do have a memory, e.g., if there are *wear factors* which lead to higher failure rates later in time, the *Weibull* family of curves are usually used to describe the failure density functions. These are of the form

$$f\,(t;\alpha,\beta) \;=\; \alpha\,\beta\,t^{\beta-1}\,e^{-\alpha t^{\beta}} \qquad (2.47)$$

where $\beta = 1,2,3,\ldots..$

Notice that $\int_0^\infty f\,(t;\alpha,\beta) \;=\; 1$

and that β=1 gives the exponential distribution. A few of the Weibull curves are shown in Fig. 2.5.

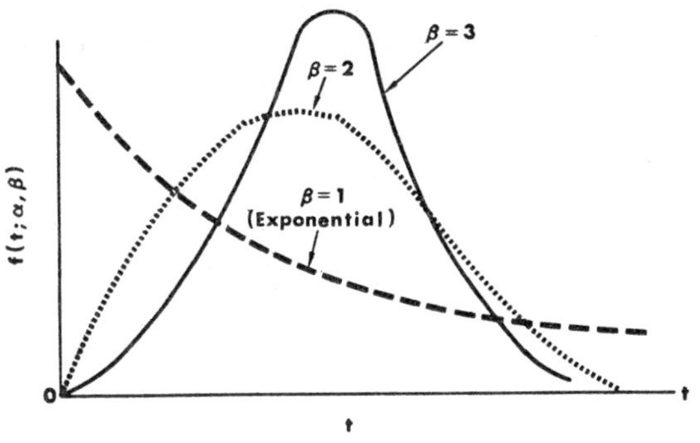

Fig. 2.5. Weibull Density Functions.

An important part of reliability theory is *reliability estimation*. If we test a reasonable sized representative sample of components and find that the fraction of good samples is $\overline{x}$ then $\overline{x}$ is clearly our best estimate of the fraction of good samples in the batch, but we usually want to know the confidence interval. That is, we want to be able to state with some small probability of error α that the reliability of the batch is at least as good as some number R. Clearly the smaller the value of α and the larger the value of R, the greater the number of samples which we must test.

The mathematical problem can best be visualized in the following way. Suppose the required confidence level is 99% (1% probability of error will be accepted). We can reject all values of reliability so low that with 99% probability they would have given us more failures than we observed.

The specifiable reliability, therefore, is that which has 99% probability of resulting in more than the observed number of failures, i.e., 1% probability of resulting in less failures. Thus, from the binomial distribution (Eq.2.1) this minimum reliability is that which satisfies

$$\sum_{k=0}^{k=F} \frac{n!}{k!(n-k)!}\,(1-R)^k\,R^{n-k} = \alpha$$

where: R is the probability of a successful test
1-R is the probability of a failure
F is the number of failures
n is the number of samples tested.

To transform this equation into the more usual form, let n-k=j. Then k=n-j and the summation is from j=n-F to j=n. n-F can be written as $n\overline{x}$ where $\overline{x}$ is the observed success rate. Thus

$$\sum_{j=n\overline{x}}^{n} \frac{n!}{j!(n-j)!} (1-R)^{n-j} R^{j} = \alpha \qquad (2.48)$$

where n = number of samples tested

$\overline{x}$ = average success rate among the samples

R = reliability of the batch

α = the small probability of error which we are willing to accept.

Notice that if $\overline{x} = 1$; no failures among our samples, Eq.2.48 reduces to $R^n = \alpha$. So that if we want a large R and a small α, we will still need to test a large number of samples. If we observe one or more failures, we must use Eq.2.48 and the binomial distribution tables to find our answer which, of course, will be a larger number of samples. This equation is the basis for *sequential testing procedures* where we start with an initial sample size which will be adequate assuming that we observe no failures, and add more samples if we observe failures.

If the reliability is related to TTF we may test a group of samples by observing their TTF and from that, estimate the MTTF of the batch and confidence limits. The best estimate of MTTF is, as one might expect, the MTTF of the samples tested. However, the establishment of a confidence interval is somewhat more complex.

Like the case of reliability discussed above, we generally wish to establish, on the basis of our sample test, with some small probability of error α, that the MTTF of the batch is greater than some number $1/\theta_o$ (see Eq.2.46). The answer is given by chi squared distribution tables, e.g., those in Ref.4, in the following way.

One can enter a *chi squared statistical distribution table* with two numbers; one of which is an integer called the number of degrees of freedom, the other is a variable with a range between zero and 1, and extract a positive value called χ^2. The relationship between these quantities and the MTTF confidence limit problem is

$$P\,(2\,n\,\theta\,\overline{T} \leqq \chi^2_{\alpha;2n}) = 1 - \alpha \qquad (2.49)$$

where P = indication of the probability of the event in brackets

n = number of samples tested

θ = inverse of the MTTF of the batch (see Eq.2.46)

$\overline{T}$ = MTTF of the samples tested

$\chi^2_{\alpha;2n}$ = value of chi squared entered in the table for the variable α and the number of degrees of freedom 2n

α = small acceptable probability that the MTTF of the batch is less than $1/\theta$.

The proof of this relationship is too complex to be demonstrated here, however the following example of the use of Eq.2.49 is presented. Suppose we test 10 representative samples and find a MTTF of 2 days, what minimum MTTF of the batch is indicated with 99% probability. Here $\alpha = .01$, $2n = 20$, χ^2, from the table, is 37.6. This means $\theta \leq 37.6/2n\overline{T} = .81$. Thus at a confidence level of 99%, we can state that the MTTF of the batch is at least 1/.81 = 1.1 days.

Further, if we wish to find the reliability of a sample of this batch in a 4 hour (1/6 of a day) mission, we can use Eq.2.45 (assuming the exponential distribution is valid) to find

$$R(1/6) = e^{-.81/6} = e^{-.135} = .874$$

Thus we can state with 99% confidence that the probability of the component surviving a 4 hour mission is at least .874.

2.8 Telephone Channel Usage.

In normal use, the demand for a telephone channel link can be expected to vary. Aside from the systematic and predictable variation with time of day, or week or year, there are statistical variations which must be accounted for in the design. If we design a link for the peak daily load which occurs at a particular time, we must recognize that there will be statistical fluctuations in that load and that on some days, unless our link is grossly overdesigned, the demand will exceed the capacity. We generally accept some small probability that the capacity will be inadequate to peak hours and provide some means for delaying calls (e.g., by not promptly providing a dial tone) if that occurs. However, we must know the probability of such an occurrence as a function of the link capacity and the average message rate.

The problem is basically quite similar to the shot noise problem. There are, in general, a large number of telephones from which a call could be placed through that link and a small probability that such a call will be placed at any particular time. Further, barring the occurrence of general telephone stimuli such as earthquakes which we cannot design for, the probability of different numbers of calls being placed is represented by a binomial distribution. Usually it can be represented by the limiting case of such a distribution with large n and small p. Thus we can take from the shot noise results the fact that

$$\overline{C} = np$$

$$V(C) = \sqrt{np} = \sqrt{\overline{C}}$$

where $\overline{C}$ = average number of calls during that peak hour

n = number of phones which could make such calls

p = probability that any particular phone will enter a call at that time on that link

V(C) = variance in C.

It is convenient at this time to utilize the *Poisson* distribution. This is the limiting case of the binomial distribution for large n and small p, but expressed in terms of its mean value which, since it is a binomial distribution, is np. The Poisson distribution is

$$P(N) = e^{-m} m^N / N! \qquad (2.50)$$

where P(N) - probability of occurrence of a particular value of N

N = number of calls at any time

m = mean or expected value of N (np)

Tables of Poisson distribution and of summed Poisson distributions are carried in most statistical handbooks, e.g., Ref. 4. Fortunately, to find an answer in terms of Eq.2.50, we do not need to know p or n, but merely their product, which is the average number of calls occurring during the peak hour. This is directly measurable.

Thus suppose the mean number of calls on a particular link during peak hours is 10 and the link can handle 15, what is the probability that the capacity will be exceeded. The table gives

$$\sum_{N=15}^{N=\infty} e^{-m} m^N / N! = .083 \quad \text{for } m = 10.$$

Therefore there is 8.3% probability that the capacity will be exceeded.

For large N and m one will usually run off the available tables of Poisson distributions. In such cases the DeMoivre-LaPlace limit theorum is generally valid, and the Poisson distribution can be replaced by the Gaussian distribution as indicated by Eq.2.15.

CHAPTER III

SIGNAL PROCESSING AND DETECTION

3.1 Detection and False Alarm Opportunities.

In Section 2.1 it was noted that the time interval between independent samples which can be drawn from noise is $1/2B_n$ where B_n is the noise bandwidth. This implies that the number of independent samples per second is $2B_n$ and we could transmit data at that bit rate. This is, of course, not a precise value because the statistical independence is a matter of degree. Further, when a signal is detected (rectified) some of the information content is lost, e.g., phase relations and polarity. Therefore it is usual to assume that the *usable number of independent samples per second* in a signal of bandwidth B is B.

This implies that if the information rate matches the bandwidth the possible number of detections per second is of the order of B, and if the signal is rarely present,

$$N_{fa} = B\,P_{fa} \tag{3.1}$$

where N_{fa} = number of false alarms per second

B = bandwidth in hertz

P_{fa} = probability of a false alarm per opportunity

By turning the signal off and on in a coded fashion we can communicate, in this bandwidth, B bits of information per second. Note however, that if the signal is present half the time, which is usual in communication circuits, there are only B/2 opportunities for false alarms since we cannot have a false alarm when the signal is present. Similarly there are B/2 opportunities for missed detections. However the information rate may be less than that indicated by the bandwidth e.g., incoherent processing may be used as discussed in Section 3.2 below. In that case the opportunities for false alarms and missed detections will be less.

3.2 Incoherent Signal Processing.

Suppose that in this bandwidth B we wish to transmit less than B bits of information per second, say B/2. We now have an opportunity for incoherent signal processing. We hold each bit of information i.e., each signal, or interval during which the signal was absent, for a period of time 2B, and adopt some decision rules e.g., requiring that the signal be detected twice in two successive intervals, before being called a detection.

Any such detection scheme is mathematically equivalent to performing some averaging or adding at the output of the detector before the signal is passed on to the threshold or decision device. Consider what would happen if we took two successive independent samples from the output of the detector and added them together. We would have a new distribution with the mean and the variance both equal to twice the value of the individual samples.# The standard deviation therefore would be $\sqrt{2}$ times as large. Notice, however, that the signal contribution has doubled while the standard deviation has increased by $\sqrt{2}$ so that our detection capability is improved.

The situation is quite analogous to the detection problem which we discussed in Section 2.1. We now have two independent samples and are better able to detect a small variation in the mean. However, the improvement in signal to noise ratio is only as the square root of the number of samples. The general result can be expressed as

$$\left(\frac{S}{N}\right)_{out} = \left(\frac{S}{N}\right)_{in} \sqrt{Bt} \quad \text{incoherent processing} \qquad (3.2)$$

where the square root of the *bandwidth time product* is the square root of the number of independent samples on which the decision is based.

This type of processing gain, called *incoherent processing gain,* is generally characterized by operations on the signal after it has been detected (rectified). Its most common form is simply a resistance capacitance circuit following the detector to smooth out the fluctuations in detector output and t is simply the RC time constant. In order to accomplish this type of processing gain, one needs to know very little about the signal. In fact, all we need to know is that the signal is within the bandwidth and that, if present, *it persists for some length of time t.* Its frequency can wander around within the band with no consequences, but if it does not persist for the length of time over which we are summing, then there is nothing to be gained in the summation and in fact we will lose signal to noise ratio.

3.3 Coherent Signal Processing.

Coherent processing generally involves operating on the signal ahead of the detector. The simplest form of coherent processing; in fact it is so simple it would probably not be dignified by that name, can be used with a signal with adequately steady frequency. It consists of just narrowing the bandwidth, which effectively reduces the noise power into the detector. However, again we require some minimum signal persistance because a signal which persists for the length of time t will be significantly attenuated in passing through a filter whose bandwidth is less than 1/t. This phenomenon can be conceptually understood from a number of different points of view, of which probably the most direct is to recognize that the frequency content of a pulse of length t is spread over a minimum bandwidth of 1/t. If the bandwidth of the filter or amplifier through which we pass the pulse is less than this, part of the energy will be lost.

By reducing the bandwidth by this additional filter (whose width we will call B_f), we have reduced the noise power to the detector by the ratio B/B_f. Since the best we can do in narrowing the filter is to make B_f equal to 1/t, where t is the duration of the signal, the best processing gain which we can achieve is

$$(S/N)_{out} = (S/N)_{in} \; 2B\,t \quad \text{coherent processing} \qquad (3.3)$$

where the factor 2 which we discarded in Eq.3.1 is put back.##

#See, for example, Eq.2.6.

##It is possible to show by a detailed analysis that the factor of 2 does in fact belong in Eq.3.3 but not in Eq.3.1. However, these detailed derivations will not be presented here. The reader is referred to Ref. 13.

Coherent processing, however, is usually accomplished by much more sophisticated means. Suppose, for example, that the signal we are looking for is swept cyclically in frequency across the band of our amplifier. If we know enough about the signal we could follow that sweep with a local oscillator so that the i.f. frequency would be constant. We could then narrow the bandwidth of the i.f. Alternatively we could, assuming sufficient knowledge of the signal, correlate it with its replica. This process is essentially one of multiplying the input by a replica of the signal we are looking for and integrating the product over the expected duration of the signal. If the signal and its replica are perfectly in step, then the product will always be positive and the integral will be large.

This correlator technique is one particular embodiment of what is called in the literature a "*matched filter*". A matched filter is defined as a network whose frequency response function maximizes the output peak signal to mean noise power ratio. It is an optimum method for the detection of signals and noise. It is easy to define, but usually very difficult to do or to even approximate except for signals of specific wave form.#

No matter what embodiment a matched filter may take, it must have some integration or averaging time inherent in it, that averaging time cannot be longer than the duration of the signal, and *its performance cannot exceed that indicated by Eq.3.3.*

Using Eq.3.3 we can express the performance of a matched filter or other completely coherent detector as

$$\left(\frac{S}{N}\right)_{out} = \frac{S_{in}\,t}{N_{in}/B} = \frac{E}{N_0} \qquad (3.4)$$

where E = total energy content in the signal pulse to be detected

N_0 = input noise power per unit of bandwidth.

Notice that in Eq.3.4 the signal to noise output, which will go to our detector and threshold device, is quite independent of any characteristic of the receiver, even its bandwidth. This is as it should be, since we have defined the receiver as optimum. However, it also indicates that coherent signal processing or matched filters contain no magic, since there is a strict upper limit to what can be accomplished which depends upon the total energy in the signal pulse to be detected, and the background noise intensity. If we have a coherent detector we can trade off, in transmission, signal power against signal duration so long as we retain the same energy in each signal pulse. To do this, however, *we have to know all about the signal* in detail, and only ask the question: "Does the signal exist or does it not exist?" If there is any uncertainty about the shape, the phasing, or any other characteristics of the signal, then the best we can achieve in signal processing is somewhat less than what can be accomplished with a matched filter.

In cases where the bandwidth time product is very large, the difference between coherent and incoherent processing gain can be very significant. However, the designer generally has some latitude in the system characteristics, and the important application of large coherent processing gain usually occurs in data links of very low information content and very low signal level where the peak power which can be utilized is limited by available components; for example, in transmitting data from space vehicles.

In considering applications of this type it is important to note that we now have in practical use frequency generators with stabilities of the order of 1 part in 10^{10}, that is 1 microsecond in 2.8 hours. This presents important opportunities for coherent detection with very large processing gains. Frequency stability is required in such applications because of the narrow frequency channels required or the need to replicate the signal internally for processing at precisely the same rate at which the signal arrives. Of

---

#See, for example, Reference 11.

course that does not mean that we can always work to those limits because in any particular situation there are other factors which destroy our ability to do this, i.e., unpredictable relative motion of source and receiver, phase shifts in the atmosphere and other media, etc.

3.4 ROC Curves and the Calculation of Error Rates.

We are now technically in a position to calculate for any simple receiver system operating with a specified signal to noise ratio, a relationship between detection probability, false alarm rate, and decision threshold settings.

For a linear detector, for example, we could start with the signal to noise ratio into the receiver, apply whatever processing gain is utilized or will be utilized in accordance with Eq.3.2 or 3.3 to obtain the improved signal to noise ratio, estimate the decision threshold, apply Eq.2.18 to determine the false alarm probability, and Eq.2.34 to determine the detection probability. To determine the false alarm rate we divide the false alarm probability by the signal pulse duration t, the same t which appears in the processing gain equations, and multiply by the fraction of the time no signal is present. We could then readjust the decision threshold and repeat, etc.

However, the use of Eq.2.18 and 2.34 is difficult. We would be required to integrate over the density functions to find the probabilities of detection and false alarm. Actually, engineers do not do this type of calculation in their normal work, but rely upon previously plotted curves such as Fig.2.7 in Ref.14. Using curves such as that, which are available in many texts, one can readily determine the false alarm probability and detection probability for a given threshold and signal to noise ratio (after processing gain). We then apply the same time factor to determine false alarm rates.

An alternative, still more accurate method, and one which has come into increased popularity in the past few years is the method based upon receiver operating characteristics (ROC) curves. This method was developed by Peterson and Birdsall (Ref.15). It is based upon the characteristic curves shown in Fig. 3.1.

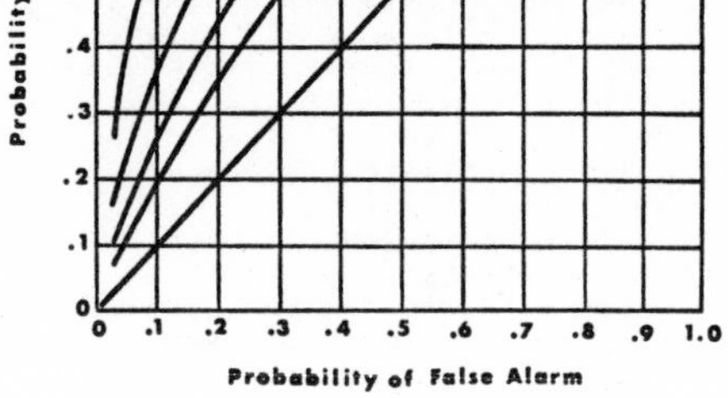

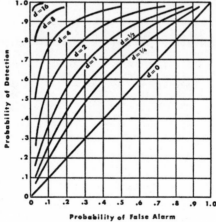

Fig. 3.1. ROC curves, linear scale.

In Fig.3.1 d is the *detection index* (a ratio) which is given by

$$d = \frac{\left(M_{(S+N)} - M_N\right)^2}{\sigma^2} = (S/N)_T \qquad (3.5)$$

where $M_{(S+N)}$ = average value of the signal plus noise envelope

M_N = average value of the noise envelope

σ^2 = variance in the noise envelope

$(S/N)_T$ = signal to noise ratio at the threshold device.

In the case where there is no signal processing involved, d is simply the *input signal to noise power ratio* (not in dB). However, in the more general case the signal processing can be included in the calculation of d as follows. For coherent processing gain

$$d = 2\ B\ t\ S/N \tag{3.6}$$

While for incoherent processing gain d is calculated from

$$d = B\ T\ (S/N)^2 \qquad (\text{if } S/N << 1) \tag{3.7}$$

$$d = \sqrt{B\ T}\ S/N \qquad (\text{if } S/N > 1) \tag{3.8}$$

where S/N = signal to noise ratio before processing.

The coherent processing gain is plainly visible in Eq.3.6 as twice the bandwidth time product. The incoherent processing gain is similarly visable in Eq.3.8. However, the incoherent processing gain is not readily recognizable in the Eq.3.7. To find something which looks like it we take the square root of Eq.3.7.

$$\sqrt{d} = \sqrt{B\ t}\ \ S/N \tag{3.9}$$

Since d is generally larger then one in a real system $\sqrt{d} < d$ and for very small signal to noise ratios the Peterson-Birdsall detection index is less than what one would calculate using incoherent processing gain. The difference can be significant, and when it is the Peterson-Birdsall method given here is more accurate than the use of the incoherent processing gain formula.

For convenience in using this method, we have reproduced as Fig.3.2, on a probability scale, the same ROC curves which are shown in Fig.3.1.

3.5 Approximate False Alarm Probability.

We present here a useful approximation by means of which one can estimate what error probability to expect in a detection circuit as a function of the signal to noise ratio (after processing) with which the circuit must work. The treatment is approximate because we do not properly treat the probability density function which exists when the signal is present. The approximation is useful because it does not require the use of charts and tables and is simple to remember and use.

The circuit which we analyze is simply a linear detector (rectifier) followed by a threshold circuit. When the detected output power exceeds the threshold, the receiver will indicate that a signal is present. When the detector output is less than the threshold it will indicate that no signal is present. We will consider, because it is easier to analyze, that the decision circuit responds to the power out of the linear detector. If the signal is present, without noise, the voltage out of the detector will be the peak signal voltage and the available power will be

$$\overline{W} = 2S/R \tag{3.9}$$

where S = the signal power

R = 4ρ (4 times the video output resistance of the detector as in Eq.2.23).

Next note from Eq.2.24 that with noise alone present, the average available detector output power is

$$\overline{W} = 2\sigma^2/R, \tag{3.10}$$

where σ^2 = the noise power into the detector.

We will select as a reasonable decision criterion the logarithmic mean of these two values. Thus, the decision threshold power will be

$$W_T = \frac{2}{R}\sigma\sqrt{S} \tag{3.11}$$

The probability of the noise alone exceeding the threshold, from Eq.2.23 is then

$$P(W > W_T) = \frac{R}{2\sigma^2}\int_{W_T}^{\infty} \exp - [\frac{WR}{2\sigma^2}]\, dW$$

$$= - \exp - \frac{WR}{2\sigma^2}\Big]_{W_T}^{\infty} = \exp - \frac{W_T R}{2\sigma^2}$$

Substituting W_T from Eq.3.11 we find

$$P_{FA} = \exp - \sqrt{S/\sigma^2} = \exp - \sqrt{\eta} \tag{3.12}$$

where η = signal to noise power ratio. (not in dB)

The method of choosing the threshold assures that the missed detection probability will be roughly comparable to the false alarm probability, and the discussion in Section 2.5 assures that the result is substantially independent of detector type.

Equation 3.12 can be used to obtain an estimate of the attainable error probability from the signal to noise ratio. Thus, for example, if we are working to a signal to noise ration of 12 dB, $\sqrt{S/N}$ = 6 dB = 4, and the attainable false alarm probability is of the order of e^{-4} = .022 or 2.2%. The missed detection probability will be about the same.

If we compare this result with Fig.3.2 (12dB corresponds to d = 16) we will find fairly close agreement. The figure indicates that with d = 16 and P_{FA} = 2%, P_{MD} = 3%.

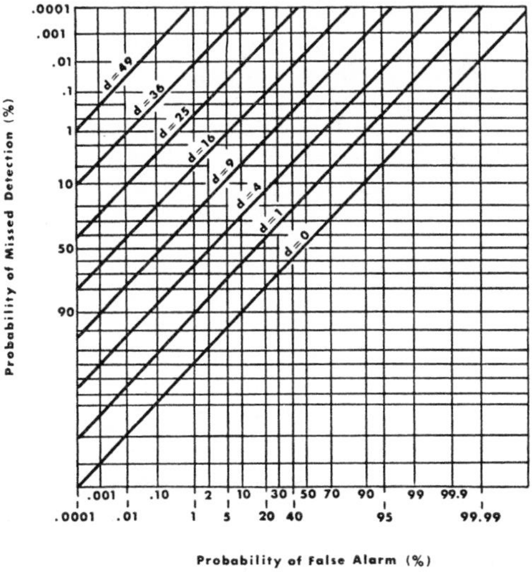

Fig. 3.2. ROC curves, probability scale.

3.6 Parity Checks.

An important technique used for reducing errors in binary digital systems is the parity check. So far we have considered that signal decisions may be simply right or wrong. This technique introduces a third category - *correctness in doubt,* which calls for a repeat of the message. The technique consists simply of adding an additional bit to each message which may be either a 1 or a zero; selected so that the total number of ones in the message is an even number. (An odd total number of ones may also be used in exactly the same way if that is the pre-arranged parity.) The importance of the technique is that if one error is made anywhere in the group of digits, the parity will be wrong (i.e., the number of ones will be odd instead of even or vice versa) and the error will be recognized at the receiving end. The receiver, of course, will not be able to recover the correct message because it will not know which of the digits was wrong, but it can request a repeat of the message. Thus the errors will go undetected only if two are made in the same message.

Consider a case where we are working with a detection index of 20. Fig. 3.2 indicates that we can achieve a 99% probability of detection with a 1% false alarm probability. Suppose we have 5 bit messages to transmit and we do not use parity checks. The overall probability of an error in the message is about 1% x 5 = 5%. On the other hand, if we add a 6th bit for a parity check, the single error case will be detected, but an undetected error will occur if two errors are present in a single message. The probability of this can be computed by the binomial distribution (Eq.2.1) with

p = probability of an error per bit (.01)
$q = .99$
$n = 6$
$x = 2$

Thus,

$$P^6_{(2)} = \frac{6!}{2!\,4!}(.01)^2 \times (.99)^4 \cong .0015$$

which is very much less than .05. Four errors in a single message will also go undetected but this is less likely by two more orders of magnitude.

The parity check, beside the complexity of the circuit, occupies additional channel capacity in two ways. First there is a 6th bit of information to be transmitted which requires time; and additional time is required to retransmit those of the messages which were in error, but the advantage can be quite great, particularly in the case of non-Gaussian noise (occasional bursts of interference). Parity check is frequently used when data is handled by computers or computer links. It has an additional advantage in that it gives a measure of the number of errors which are being made, which is valuable as a check on the system. With a parity check the percentage of undetected errors is roughly the square of the percentage of detected errors, so if the number of detected errors rises above some pre-set threshold, the system can be arranged to shut down and call for help, indicating that something is wrong in the data transmission circuits.

CHAPTER IV

ANTENNAS

4.1 Definition of Terms.

Since we will be talking about antennas which are matched to transmission lines, it is important to understand what we mean by *impedance match* in the case of an antenna. An antenna is both a receiver and a transmitter, but its impedance can only be measured by using it as a transmitter. Physically the process is simply to connect it to a transmission line, through a slotted line (which is matched to the transmission line), and transmit power through the slotted line and the antenna. We can then measure the standing wave ratio on the slotted line and if the ratio is 1 (no reflected power), we say that the antenna is matched. That means that all the energy in the transmission line which is incident on the antenna is either absorbed in the antenna or radiated. In the discussion which follows, for the most part we will be working at the level of approximation which ignores antenna losses, and consider that all the energy is radiated.

One important property of an antenna is its *effective area*# in a particular direction. This property describes its characteristics when used as a receiver. It can be defined in the following way. Consider a plane wave incident on an antenna from a particular direction with a certain energy density in watts per square meter. If the antenna is matched to the transmission line, as defined in the previous paragraph, and a load is also matched to the line, it will deliver to that load some amount of power which it extracts from the incident plane wave. If we divide that amount of power by the power density per square meter in the incident plane wave, the result is dimensionally an area in square meters. This is the effective area of the antenna in that direction at that frequency. It is perhaps helpful to visualize the antenna as chopping out all the energy passing through some area about it and gathering it in. It is important to notice that this area is a function of the direction from which the plane wave is incident. When we speak of an effective area of an antenna without indicating direction, we mean the area in the *principal direction* of the antenna, that is, the direction in which the area is a maximum and the direction in which the antenna is designed to look.

An *omnidirectional antenna* is one which radiates uniformly in all directions over the full 4π solid angle about it. We know of no way to build such an antenna, but it is a convenient concept for use in reasoning and analysis.

The *gain* of an antenna is a characteristic which defines its transmitting characteristics. The gain in any direction is defined with reference to an omnidirectional antenna. It is the ratio of the power radiated per unit solid angle in that direction to the power which would be radiated by a matched omnidirectional antenna with the same power coupled into it; usually expressed in dB. As with cross sectional area, when we speak of the gain of an antenna without any directional reference we mean the gain in the principal direction or the direction in which the antenna is designed to transmit.

#Sometimes called effective aperture.

4.2 Basic Properties.

The concepts of noise and the second law of thermodynamics which we used in Chapter I to prove certain properties of filters and mismatches are also useful to derive certain characteristics of antennas. Consider a transmission line at one end of which there is a matched resistor and at the other end a matched lossless antenna. Consider that the entire combination is in a space completely enclosed by some surface which is opaque to microwave radiation, and that the resistor, antenna and enclosed space are all at a uniform temperature T_1. The resistor is coupling an amount of power equal to kT_1 per unit of bandwidth into the transmission line and through it to the antenna where, because the antenna is matched, it is radiated. If the second law of thermodynamics is not to be defeated, the space around the antenna must return to that antenna and resistor exactly the same amount of power.

Let us, at this point, accept the fact, which we will prove later, that under such conditions the radiation flux of energy in the enclosed space will be omnidirectional. That is, it will be the same when viewed in any direction from the inside of the space with no predominance of energy flux from any one direction over that from any other direction. In this case we can designate the flux of energy per steradian per unit bandwidth in any direction as $\Phi(T_1)$. We can therefore write

$$kT_1B = B\int_0^{2\pi}\int_0^{\pi} \Phi\,(T_1)\,A_e\,(\theta,\phi)\,\sin\theta d\theta d\phi$$

$$kT_1 = \phi(T_1)\int_0^{2\pi}\int_0^{\pi} A_e\,(\theta,\phi)\,\sin\theta d\theta d\phi \qquad (4.1)$$

where $A_e(\theta,\phi)$ = effective area of the antenna in the θ,ϕ direction.

This must be true for any antenna.

Therefore, the integral of the cross sectional area of an antenna over all directions has some fixed value which may depend upon frequency since $\Phi(T_1)$ may depend upon frequency, but it is independent of the antenna so long as it is matched. This is a rather surprising result. We are saying that *a very large antenna and a very small antenna each placed in an isotropic radiation field will collect exactly the same amount of energy*. The large antenna will collect more in some particular direction in which it is pointed, but it will collect less in other directions and quantitatively, these must compensate. Thus, if an antenna has a very large cross sectional area in some particular direction, its receiving pattern must be very narrow because its receiving pattern integrated over a complete sphere cannot be greater than the result of a similar integration performed for an omnidirectional antenna or any other antenna. Thus the integral over all directions of the effective area of any antenna is a function only of frequency. Although we will not demonstrate it here because it requires some complex thermodynamics, this function = $(c/f)^2$ or λ^2. That is

$$\int_0^{2\pi}\int_0^{\pi} A_e(\theta,\phi)\,\sin\theta d\theta d\phi = \lambda^2 \qquad (4.2)$$

where A_e = effective cross sectional area of any antenna

λ = wavelength of the radiation.

This would seem to imply (from Eq.4.1) that $\phi(T)$ or the thermal equilibrium radiation per steradian is kT/λ^2, but it is really twice this value ($2kT/\lambda^2$) because of polarization (Ref.15). The antenna can receive only one polarization, but the radiation field contains two. Even a circularly polarized antenna can receive only half the energy incident upon it from a noise field because it must be either a right hand or left hand circularly polarized antenna, and whichever it is; the alternative constitutes an orthogonal mode which cannot be absorbed.

Next we shall prove by a similar thought experiment, our previous assumption that the thermal radiation field inside any kind of an opaque surface at uniform temperature is isotropic. We assumed this to get Eq.4.1. To prove it, merely consider two highly directional antennas each connected to a transmission line and a resistor; placed in such a thermal noise field but looking in different directions. As before, the resistors, transmission lines, and attenuators are also at temperature T_1. If the radiation field is not isotropic, these two antennas will collect different amounts of energy and their associated resistors will end up at different temperatures. This, of course, would violate the second law of thermodynamics, so the field must, in fact, be isotropic.

Next we can use the same approach to prove that the radiation pattern of any antenna must match its cross sectional area pattern. That is, if we were to plot the cross sectional area of an antenna in any direction as a function of that direction and plot the gain in any direction as a function of that direction; the two plots for the same antenna must be identical in shape. To prove this, consider the antenna placed in an isotropic thermal noise field and connect it to a transmission line and a resistor as in the two previous cases. If the antenna tended to absorb more energy than it emitted in some direction and emit more energy than it absorbed in another direction, it would be effectively taking energy which arrives at it from one direction and emitting it in another direction. This would destroy the isotropic character of the radiation field which we proved in the previous paragraph to be a necessity if we are to preserve the second law of thermodynamics. *Therefore the receiving and transmitting patterns must be the same for any antenna.*

One may note that if a reflecting surface is placed in such a isotropic thermal radiated field, it does, in fact, take energy incident from one direction and send it into another. However, this does not destroy the isotropic character of the field because such a surface has reflective symmetry. That is, there is a reverse phenomenon with energy traveling along exactly the same path in the opposite direction. For the antenna there is no such compensating reverse path.

The concept discussed here has an important parallel in physical radiation theory. At infrared, for example, the emissivity of a surface must be proportional to its absorbtivity. If we had a surface which was a good absorber but a poor emitter, a body covered with such a surface would become warmer than ambient when placed in an isothermal radiation field. The second law of thermodyanmics tells us that this cannot happen and therefore such a surface cannot exist. The analogy to the material presented in Section 1.1 is also clear.

It remains therefore only to determine the constant of proportionality between the cross section in any particular direction and the area in that direction.

The power absorbed by an antenna in the θ,ϕ direction is $A_e(\theta,\phi)kT/\lambda^2$ per steradian in a thermal equilibrium radiation field, and the power radiated by it, assuming it is connected to a transmission line and matched resistor is $G(\theta,\phi)\ kT/4\pi$ per steradian (where $kT/4\pi$ is the power per steradian radiated by an isotropic antenna). Since these two must be equal if the antenna is to radiate as much power as it receives in every direction, the antenna gain in any direction must be related to the effective area in that direction by

$$G(\theta,\phi) = 4\pi\ A_e\ (\theta,\phi)/\lambda^2 \qquad (4.3)$$ m

Equation 4.3 indicates that *the effective area of an isotropic antenna is* $\lambda^2/4\pi$. This is easy to remember because it is the area of a circle whose circumference is 1 wave length. Thus, a matched isotropic antenna would collect from a plane wave radiation field, all the power which is within a circle of circumference λ.

4.3 Power Transfer.

The power transfer between two antennas in the absence of any absorption in the medium between them is given by

$$P_r = P_t G_t A_e / 4\pi r^2 \tag{4.4}$$

where P_r = received power

P_t = transmitted power

G_t = gain of the transmitting antenna in the direction of the receiver

A_e = effective area of the receiving antenna in the direction of the transmitter

r = distance between them (same units as A_e).

Of course, any losses in the antenna feeds must be added separately since they are not included in Eq.4.4.

For large antennas of the general configuration of horns or parabolic reflectors, the relationship between the actual area of the antenna and the effective area, (in the principal direction) can be described in terms of an *aperture efficiency* as given by Eq.4.5.

$$A_e = \rho_a A \tag{4.5}$$

where ρ_a = aperture efficiency

A = physical area of the antenna.

Aperture efficiencies of large antennas vary between about 50% and 90%. It is important to note that there is, in antenna design, a tradeoff between aperture efficiency and side lobe suppression. To understand this tradeoff it is easiest conceptually to think of the antenna transmitting, using the relationship between gain and effective area given by Eq.4.3. To achieve good side lobe suppression, it is necessary (as discussed in Section 4.6) to taper the illumination of the main aperture, that is of the horn or parabolic reflector, so that the edges are not strongly illuminated.

Having tapered the illumination, it is clear that the edges of the antenna do not contribute as much to the gain of the antenna in its principal direction as they might otherwise contribute. Thus the gain and hence the effective area of the antenna will be smaller. Therefore aperture efficiency is not necessarily a measure of antenna excellence (Ref.16).

The effective area of small antennas (small area compared to λ^2) does not bear any significant relationship to physical area. Thus, for example, for whip antennas or loop antennas or any antenna which does not have high gain, the effective area is not related to any of the physical dimensions of the antenna. It is interesting to note from Eq.4.3 that if the antenna does not have any great amount of gain, the effective area depends principally upon the frequency, and will become quite small as the frequency increases. Thus, for example, an omnidirectional antenna in the broadcast band at 1 megahertz can have a cross sectional area of the order of 10^4 square meters. At 10 gigahertz, on the other hand, the best that can be accomplished is 10^{-4} square meters. Therefore, at very high frequencies, omnidirectional antennas are of limited utility as receivers.

4.4 Noise Temperature of Antennas.

The noise temperature of an antenna is that temperature which corresponds to the thermal noise output which the antenna supplies to a matched receiver connected to it. To the approximation that the antenna is lossless, this power does not originate in the antenna itself but is collected by it from the radiation field in which it is placed. The radiation temperature of the antenna can be calculated by taking its available thermal noise output power per unit of bandwidth and dividing it by Boltzman's constant k. Using the thermal energy flux density developed in Section 4.2, (kT/λ^2 per steradian for a single polarization) we can calculate the effective temperature of the lossless antenna as

$$T_{ea} = \frac{P_{rt}}{kB} = \frac{1}{kB} \iint A_e\,(\theta\phi)\ [kB\ T_s\,(\theta,\phi)/\lambda^2]\sin\theta\ d\theta d\phi \qquad (4.6)$$

where T_{ea} = effective temperature of a lossless antenna

P_{rt} = available output noise power

$T_s(\theta\phi)$ = temperature of the radiation field reaching it from the θ,ϕ direction.

If we substitute for A_e from Eq.4.3 we have

$$T_{ea} = \frac{1}{4\pi} \iint G\,(\theta\phi)\ T_s\,(\theta,\phi)\ \sin\theta\ d\theta d\phi \qquad (4.7)$$

If T_s is the same over all directions in which the antenna has significant gain, we can take it out of Eq.4.7 as a constant, and we have

$$T_{ea} = \frac{T_s}{4\pi} \iint G\,(\theta\phi)\ \sin\theta\ d\theta d\phi \quad \text{(for constant } T_s\text{)} \qquad (4.8)$$

but it is easily shown from Eq.4.2 and 4.3 that the integral is 4π. Therefore in the case where T_s is constant in all significant directions, *the effective temperature of the antenna is the temperature of the radiation sources which it sees.*

However, if the antenna is looking at a low temperature sky with its main beam and the side lobes are looking at the ground or at the earth environment, they may easily make a disproportionately large contribution to its effective temperature because their temperatures may be of the order of 100 times that of the sky, so that a side lobe which is 30 dB down but which occupies a solid angle which is 10 times that of the main beam, may dominate in determining the effective temperature of the antenna. Equation 4.7 will, of course, properly account for all side lobes if the gain which appears in that equation includes side lobes. *The fraction of the radiated power of an antenna which is in its main beam is called the main beam efficiency.*

Because of this side lobe temperature effect, it is common to cover the ground around certain antennas with metallic mesh. This material is a good reflector and because of that, the temperature which the antenna sees on those side lobes which look in the direction of the mesh is not earth temperature but reflected sky temperature. The radiation received is that reflected from the sky.

4.5 Antenna Losses.

Thus far we have treated the antennas as lossless. However, any actual antenna has significant losses in its feed structure and may even have significant skin losses in the structure of the antenna itself. These losses can be accounted for in a very straightforward way by considering the antenna as a lossless antenna whose output is coupled through an attenuator. This attenuator, of course, is at the physical temperature of the antenna and not at its radiation temperature. Therefore, these losses are important not merely because they reduce the signal, but also because, particularly when the effective temperature of the lossless antenna is low, they may add very significantly to the thermal noise output of the antenna.

A very common type of antenna is one which employs a parabolic dish fed by a horn at its focus. Such an antenna is apt to have rather large feed losses because of the necessity for tradeoffs between feed losses and *aperture blockage*. That is, the wave guide leading to the feed horn must be small in order to minimize its interference with the pattern of the parabolic dish; and this smallness leads to large wave guide losses. An important partial solution to this is the *Cassegrain antenna* (Ref.17) in which the feed horn is in the surface of the parabola and projects the energy forward to a small hyperbolic subreflector which reflects it back into the parabolic reflector. Supports are required for the hyperbolic subreflector but the transmission to it and back to the main reflector are effectively lossless. The arrangement is shown schematically in Fig.4.1.

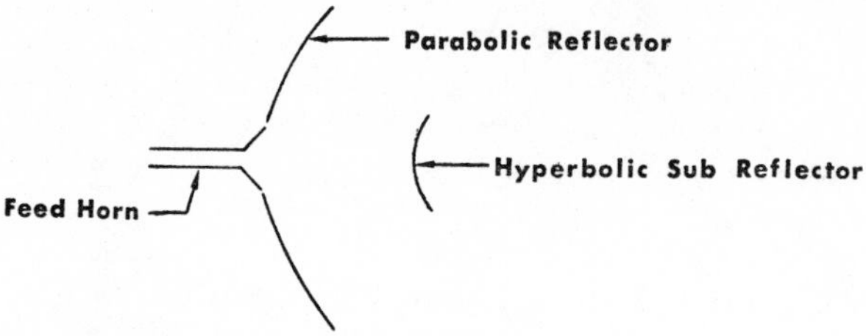

Fig. 4.1. Cassegrain Antenna.

An alternative arrangement which has been adopted and developed by the Bell Telephone Company as part of its TL radio relay system, is shown in Fig. 4.2. The surface is somewhat more difficult to design and fabricate, but it has the advantage of minimizing the feed losses since large wave guides can be used without aperture blockage and in general it is more convenient to bring the power to the antenna from below (Ref.18).

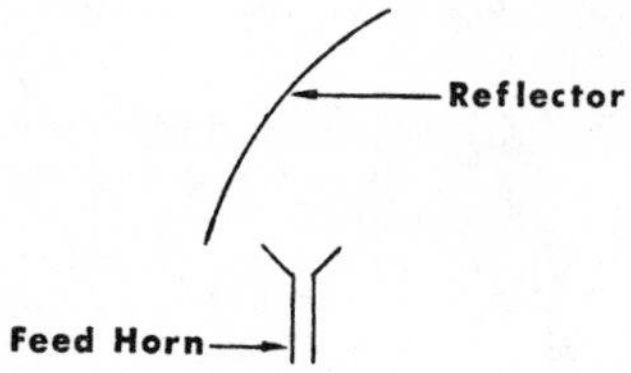

Fig. 4.2. Periscope Antenna or Offset Feed Parabolic Antenna.

4.6 Antenna Pattern Characteristics.

Antenna pattern calculations are based upon the concept of vector potential and the uniqueness theorems of boundary value problems. Basically, we are trying to find a solution to the wave equation in a space which is bounded by the antenna configuration and infinity. We assume, of course, that the fields are 0 at infinity and we know their values over the antenna. Therefore, we know the values of the fields over the complete boundary of the space of interest, and from this we are able to calculate the fields throughout the region because they are uniquely determined by these boundary values.

A further fact in connection with such boundary value problems is that only the tangential fields at the boundary are significant. That is, the fields normal to the boundary can be derived from these tangential fields so that the problem is completely specified by specifying the tangential electric and magnetic fields. If, as is usual in antenna problems, the antenna surface is a conductor, then the tangential electric field will be 0 at that boundary and there is a one to one relationship between the tangential magnetic field and the surface currents. (See Chapter V for further details.) Therefore, we can compute the antenna pattern from the currents in the dish.

The method of computation is perhaps most easily visualized as being based upon a superposition of many small current dipoles. The field of a small dipole at some distant point (sufficiently distant that near field effects vanish) is given by#

$$H_\phi = \frac{j\, I_0\, h}{2\lambda r} \sin\theta \exp(-j\, 2\pi r/\lambda) \tag{4.9}$$

$$E_\theta = \eta_0 H_\phi \tag{4.10}$$

where $j = \sqrt{-1}$

I_0 = current amplitude

h = length of the current dipole

λ = wavelength in air

r = distance from the dipole to the field point

$\eta_0 = \sqrt{\mu_0/\varepsilon_0} = 377\ \Omega$

and the dipole is oriented in the $\theta = 0$ direction

To perform the superposition in an integral sense, we replace $I_0 h$ in Eq.4.9 by $\int J_0 dA$,

where J_0 = surface current density on the antenna

dA = element of area.

We then integrate over the surface of the antenna. In this integration it is usual to ignore all but the main reflector and take care of subreflectors, horns and other obstructions by means of a blockage factor. However, we must know the distribution of current across the main reflector. It is usually not uniform. The integral will generally have the form indicated by Eq.4.11.

$$H_\phi = \frac{j}{2\lambda r_1} \sin\theta \int J_0\,(x,y,z) \exp(-j\, 2\pi r_1/\lambda)\, dx dy \tag{4.11}$$

where r_1 = distance from the point x,y,z on the antenna to the field point at r,θ,ϕ where H_ϕ is being calculated.

r_1 and $\sin\theta$ have been removed from the inside the integral in 4.11 as an approximation which is valid in the *Fraunhofer region* where the dimensions of the antenna are very small compared to the distance from the antenna to the point at which the field is being calculated, although they are not very small compared to a wavelength. The antenna is always designed so that there is some particular direction in which all the radiating elements are in phase. This is the direction of the maximum gain of the antenna and if we designate that as $\theta = 90°$ $\phi = 0$ direction, the resulting integral can generally be transformed to a plane surface integral, e.g., the surface across the parabola opening and written in the form (Ref.20):

#See, for example, Ref.19.

$$H_\phi = \frac{j}{2r\lambda} \exp -2\pi jr/\lambda \sin\theta \iint J_0 (x,y) \exp -2\pi j \left(\frac{x}{\lambda}\cos\theta + \frac{y}{\lambda}\sin\phi\right) dxdy \qquad (4.12)$$

The approximations involved in reaching Eq.4.12 are valid in the Fraunhofer region. The equation is *similar to a double Fourier transform* and has very similar characteristics. Thus the distribution of current across the aperture affects the radiated pattern in a similar way to that in which the shape of a pulse affects its frequency spectrum. If the pulse is a square one and not tapered, it will have very significant frequency side bands. Similarly, if the distribution of current across the antenna is uniform and not tapered, the side lobes of the antenna pattern will be large. However, it is this uniform distribution across the aperture which will give the maximum output of signal in the principal direction and therefore, as noted in Section 4.3, there is a tradeoff between this maximum gain or maximum aperture efficiency and side lobe suppression.

Further, if J_0 is very extensive in the x direction, giving a large gain, the pattern will be narrow, just as a pulse which is very long in time has a very narrow frequency spectrum. This general characteristic can also be derived from Eq.4.2 and 4.3.

4.7 Beam Widths and Useful Approximations.

From Eq.4.2 and 4.3, one can readily show that the integral over the complete solid angle of the gain of any antenna must be 4π. Therefore for a beam pattern which is circularly symmetric about the θ=0 axis,

$$4\pi = \iint G(\theta\phi) \sin\theta \, d\theta d\phi = 2\pi \int G(\theta) \sin\theta \, d\theta \qquad (4.13)$$

and if we consider the gain to be approximately G_{max} over some angular range from 0 to θ_1 and zero outside that region, we can write

$G_{max} = 4\pi/\Omega$ where Ω is the solid angle occupied by the beam, or from 4.13,

$$G_{max} \cos\theta \Big]_0^{\theta_1} = -2; \text{ or } 1 - \cos\theta_1 = 2/G_{max} \qquad (4.14)$$

For small θ_1, $\cos\theta_1 \equiv 1 - \theta_1^2/2$. Therefore

$$\theta_1^2 \cong 4/G_{max} \text{ in radians} \qquad (4.15)$$

This method of estimating beamwidths from gain assumes that the main beam is flat topped and steep sided and it ignores side lobes. However, it usually gives an answer within a factor of 2 of the correct beamwidth. Thus for example, if Gmax =40 dB (10,000), this approximation gives θ_1 = .02 radians or about 1.2°, and the full width of the pattern will be 2.4°. To obtain a more accurate answer we must know if the illumination of the antenna aperture is uniform or tapered, and how it is tapered. For a uniform illumination a more accurate calculation based directly on Eq.4.12 gives 1.77° between the nulls and 1.58° between half power points. For illumination in the shape of a cosine function with a maximum at the center and zero at the ends, the more accurate calculation gives 1.73° between the half power points, and a $\cos^2$ illumination gives 2.11° between the half power points.

We can also express this angular width in terms of the dimensions of a circular antenna by using the aperture efficiency (ρ_a) to relate the effective area to the actual area. In this case, and to this degree of approximation,

$$G_{max} = 4\pi A_e/\lambda^2 = 4\pi \rho_a \pi a^2/\lambda^2 = (2\pi a/\lambda)^2 \rho_a \qquad (4.16)$$

and using Eq.4.15

$$\theta_1 \cong \frac{2}{\sqrt{G_{max}}} \cong \frac{\lambda}{\pi a\sqrt{\rho_a}} \text{ radians} \qquad (4.17)$$

and the angle between the nulls on either side of the main beam is

$$2\theta_1 \cong \frac{2\lambda}{\pi a\sqrt{\rho_a}} \cong \frac{4\lambda}{\pi d\sqrt{\rho_a}} \text{ radians} \qquad (4.18)$$

where a = radius of the antenna dish and d is its diameter.

Again the precise answer depends upon the illumination pattern of the antenna. For a uniformly illuminated circular dish for example, (ρ_a = 1) the half power width of the beam is given accurately by

$$2\theta_1 = 1.03\ \lambda/d \qquad (4.19)$$

so that the approximation of Eq.4.18 is high by about 20%

The relative power gain function for a uniformly illuminated antenna is given accurately by (Ref.21)

$$F(\theta) = \left[\frac{2\ J_1\ (3.24\ \theta/\theta_1)}{3.24\ \theta/\theta_1}\right]^2 \qquad (4.20)$$

where J_1 = Bessel function of 1st kind, 1st order.

For a uniformly illuminated line array (1 dimension) the relative power gain function is given accurately by

$$f(\theta) \cong \left(\frac{\sin\ [\pi\ \theta\ d/\lambda]}{\pi\ \theta\ d/\lambda}\right)^2 \qquad (4.21)$$

CHAPTER V

PROPAGATION AND TRANSMISSION LINES

5.1 Fundamentals of Propagation.

To develop the wave equation which is fundamental to all propagation, we start with *Maxwell's equations.*#

$$\nabla \times \mathcal{H} = \sigma \mathcal{E} + j\omega\varepsilon\, \mathcal{E} = (\sigma + j\omega\varepsilon)\, \mathcal{E} \qquad (5.1)$$ m

$$\nabla \times \mathcal{E} = -j\omega\mu\, \mathcal{H} \qquad (5.2)$$ m

$$\nabla \cdot \mathcal{H} = 0 \qquad (5.3)$$ m

$$\nabla \cdot \mathcal{E} = \rho/\varepsilon \qquad (5.4)$$ m

where $\mathcal{H}$ = magnetic field vector (amperes/meter)

$\mathcal{E}$ = electric field vector (volts/meter)

σ = conductivity (mhos/meter)

ε = dielectric constant (farads/meter)

μ = permeability (henries/meters)

ρ = charge density (coulombs/meter^3)

∇ = vector operator whose x component is $\frac{\partial}{\partial x}$ etc.

$\nabla \times \mathcal{A}$ = curl of the vector A whose z component is given by the determinant

$$(\nabla \times A)_z = \begin{vmatrix} \frac{\partial}{\partial x} & \frac{\partial}{\partial y} \\ A_x & A_y \end{vmatrix}$$

and the time variation $e^{j\omega t}$ is implied in all field quantities.

As a first step in manipulating these equations, we assume that the conductivity is zero and drop σ. Note for later use that in Maxwell's equations σ only appears in the factor $(\sigma + j\omega\varepsilon)$. *Therefore we can always get any equation containing* ε *back to its complete form by substituting*

$$\left(\frac{\sigma}{j\omega} + \varepsilon \right) \text{ for } \varepsilon.$$ m

---

#Script capital letters are used to indicate vector quantities.

To get the wave equation we take the curl of both sides of Eq.5.1.

$$\nabla \times \nabla \times H = \nabla \times j\omega\varepsilon E = j\omega\varepsilon \nabla \times E$$

and substituting Eq.5.2 we have

$$\nabla \times \nabla \times H = \omega^2\mu\varepsilon H \quad (5.5)$$

Next we use the mathematical identity

$$\nabla \times \nabla \times H = \nabla\nabla \cdot H - \nabla^2 H \quad (5.6)$$

where $(\nabla^2 H)_x = \frac{\partial^2}{\partial x^2}(H_x) + \frac{\partial^2}{\partial y^2}(H_x) + \frac{\partial^2}{\partial z^2}(H_x)$, etc.

Equation 5.6 can be proved in the general case for any vector by straightforward substitution. Further, $\nabla\nabla \cdot H = 0$ since $\nabla \cdot H = 0$ by Eq.5.3. With this substitution we obtain

$$\nabla^2 H = -\omega^2\mu\varepsilon H \quad (5.7)$$

This is the wave equation. It is, in fact, a vector equation equivalent to three scalar equations one of which is

$$\nabla^2 H_x = -\omega^2\mu\varepsilon H_x \quad (5.8)$$

It is helpful to think of the wave equation, or Eq.5.8, which is one of its components in the following way. Consider that the derivatives in all directions but one, e.g., the z direction, are 0. The left hand side of the equation is then the second derivative with respect to z or the curvature. If H_x is positive, Eq.5.8 says that the curvature will be negative or downward.. If H_x is negative, the curvature will be positive or upward. Therefore, H_x *will always curve back toward zero,* and oscillate about 0 as we move in the z direction. This is the general characteristic which makes it the *wave equation.* If the sign in the equation were positive, then a positive value of H_x would result in a positive curvature and the larger H_x the more positive the curvature, so that H_x would go off to infinity in the plus or minus direction depending upon which side of 0 it started on. This is characteristic of the *hyperbolic differential equation.* It correctly describes electromagnetic fields in problems with certain types of boundaries.

If instead of being positive or negative the right hand side of Eq.5.8 were 0, the curvature would always be 0; a graph of H_x would always be a *straight line,* and this is characteristic of *LaPlace's equation* which correctly describes many static electric and magnetic fields. We will discuss LaPlace's equation in Section 5.2 below.

To be a little more explicit about the wave equation, if the derivatives with respect to x and y are 0, Eq.5.8 becomes

$$\frac{\partial^2}{\partial z^2} H_x = -\omega^2\mu\varepsilon H_x$$

whose solution is

$$H_x = H_0 \exp(\pm j\omega\sqrt{\mu\varepsilon}\, z)$$

and if we explicitly put in the time dependence we have the propagating plane wave solution

$$H_x = H_0 \exp j\omega (t \pm \sqrt{\mu\varepsilon}\, z) \tag{5.9}$$

where the phase velocity of propagation, c, can be readily recognized as $1/\sqrt{\mu\varepsilon}$ and the wavelength as $2\pi/\omega\sqrt{\mu\varepsilon} = c/f$. Both depend only on the properties of the medium. *In free space*

$$\varepsilon_0 = 10^{-9}/36\pi,\ \mu_0 = 4\pi \times 10^{-7},\ \text{and } c = 3 \times 10^8 \text{ meters/sec.} \tag{5.10}$$

Notice that our assumption that the x and y derivatives are zero is more severe than we really needed to arrive at Eq.5.9. All we really require is that

$$\frac{\partial^2 H_x}{\partial x^2} + \frac{\partial^2 H_x}{\partial y^2} = 0. \tag{5.11}$$

This is *LaPlace's equation* in the xy plane, and solutions of the wave equation which satisfy it in the plane perpendicular to the propagation direction are called TEM or transmission line waves. They have a number of special characteristics to be discussed in the following section.

5.2 LaPlace's Equation and Transmission Line (TEM) Waves.

The first two special characteristics which we have already shown are:

1. As shown in Eq.5.9, TEM waves propagate with phase velocity equal to the *wave velocity of the medium.* That is, with the velocity at which a plane wave would propagate in that medium.

2. Since the phase velocity as shown in Eq.5.9 is independent of frequency, TEM waves have *no dispersion* (in a lossless medium). This is not true, for example, of waves in a wave guide and it is not true even of TEM waves if there are losses in the medium. This can be readily seen from Eq.5.9 if we make the previously discussed substitution of $\sigma/j\omega + \varepsilon$ for ε.

The third important special characteristic of TEM waves is that the field distributions in the plane perpendicular to the direction of propagation correspond to static field distributions. That is, they are the same distributions which one would find in the static case if the fields were uniform in the direction of propagation. This can be readily seen from Eq.5.7 if we set $\omega=0$ to achieve the static solution and also hold the fields independent of z. We are then left in the x-y plane with Eq.5.11.

The last important characteristic of transmission line waves is that when the energy flows in an actual transmission line (not a plane wave), *one can speak rigorously of a voltage* across the line. We are so used to speaking of the voltage across a line that we may not realize that this is a fairly unique characteristic of TEM waves. The voltage between any two points is their potential difference or the work required to carry a unit charge between the points. If this work depends upon the path chosen between the points it is not unique and there is no definable voltage. The reason there are definable voltages between points when a TEM mode is present is because there are no longitudinal magnetic fields in such a mode. This implies, by Eq.5.2, that there is no curl to the electric field in the plane perpendicular to the transmission lines. If there is no curl in that plane, then by Stoke's Theorum the line integral of E around any closed path in the plane is 0. This is a necessary condition if we are to define a potential in that plane because then and only then the work which must be done on a unit charge to move it from one point to another is independent of the path taken between these two points. Therefore, in any transmission line carrying a TEM mode we can rigorously define the potential difference or voltage between any two points in the same transverse plane and hence in any such plane between the two conductors which are necessary to propagate a TEM mode.

On the other hand, if we talk about the voltage between two points which are not in the same transverse plane, we have a quantity which is basically meaningless and if you try to experimentally measure such a voltage you will discover the practical consequences of this fact. There is no unique measurement which one can get because the answer depends on how the voltmeter leads happen to lay relative to the conductors; just as the work done to carry a unit charge between the two points not in the same transverse plane will depend upon the path that is taken between the two points. For a more complete discussion of this subject, the reader is referred to Ref.19.

Any two conductor transmission line is inherently *balanced*. That is, at any transverse plane the current flowing in the two conductors is equal and opposite. Thus *there is a single measurable current* at any point in the transmission line. This, coupled with the fact noted above that there is a rigorously definable voltage across the line means that there is, in fact, a rigorously *definable impedance* at any transverse plane and in addition, there is a definable and measurable characteristic impedance to the transmission line.

We will utilize these facts shortly but at this point it is important to note that a coaxial line is not necessarily a balanced line since a.c. currents can flow on the outside of the outer conductor which are independent of those flowing on the inside of the same conductor. Thus there may be effectively three conductors or, viewed differently, there are actually two transmission lines involved; one inside the coax where the definable voltage is between the center conductor and the outer conductor, and one outside where the definable voltage is between the outside conductor and some distant ground plane. This multiple line configuration may become important in situations such as antenna feeds where the currents have an opportunity to leak from the enclosed transmission line to the exterior transmission line. However we normally assume that there are no currents on the line exterior and that the line is therefore balanced.

In addition, it is important to note that we sometimes speak of wave guide impedance and impedances of other types of r.f. transmission systems even though one cannot rigorously define such an impedance in terms of voltage and current. This is usually artificial and done by some special agreement to consider the ratio of E to H at some point in the transmission system as an impedance. In a plane wave, for example, E and H are everywhere uniform in magnitude, and their ratio is taken to be the *wave impedance* (377 Ω in free space).

Solutions to LaPlace's equation have a peculiar simple property which is related to the fact that *solutions in one dimension are straight lines.* (The second derivative is zero.) Consider, for example, $f(x,y)$ where f is a solution of LaPlace's equation. Let x and y be in the horizontal plane and visualize a surface where $z = f(x,y)$. Consider drawing a line in the surface $z = f(x,y)$ in the direction of increasing x and another line in the surface crossing it perpendicularly so that it is in a direction of increasing y. If one of these lines is concave upward, then the other must be concave downward. This is a direct consequence of LaPlace's equation (e.g., Eq.5.11) which says in effect that the sum of the two curvatures must be 0. Thus, the surface which we have drawn *can never have a dome or a hollow in it but only saddles.*

Thus, a solution of LaPlace's equation can never have an interior max-

imum or minimum. The maximum and minimum values must always be on the boundary. This has immediate consequences with respect to the type of boundary configurations in which transmission line waves can propagate because any field quantity which satisfies LaPlace's equation and is 0 over a complete boundary enclosing an interior, must be 0 throughout the interior. This implies that such TEM waves cannot propagate in a wave guide no matter what its shape is unless there is an additional conductor inside. Except for waves in unbounded space, all TEM waves require two conductors which are electrically isolated from each other (one may be the ground plane).

This property of solutions to LaPlace's equation forms the basis for the *relaxation method* of calculating field configurations. The basic idea is that the value of the field quantity at any interior point is the average of all the values for the points on a circle around it (or on a sphere around it in the three dimensional case). This can be most easily visualized again in the one dimensional case where the solution to LaPlace's equation is a straight line and the value at any point is the average of the two values equidistant from it on either side. The relaxation method starts with the values on the boundary and any assumed set of interior values. It then proceeds through the complete set of interior values one point at a time, changing each one to the average of the surrounding four or six points, (depending upon whether we are in two or three dimensions). When all the points have been so changed the process is repeated and it is repeated again and again until only very small changes are being made. A field configuration so reached is, in fact, a solution to LaPlace's equation. Further, given adequate definition of the boundary conditions, there are uniqueness theorums which state that there is only one solution. There are tricks to speed up the convergence and in some cases of complicated boundary configurations, the method is found to be valuable. It has the advantage that it can be readily programmed for high speed computers.

For a two conductor transmission line wave (of which TEM waves in coaxial lines are the prime example), since the voltages and current are well defined, we can also define (and directly measure by conventional techniques) an inductance and capacitance per unit length. The fact that these quantities can be defined implies that

$$\frac{dI(z)}{dz} = -j\omega C_1 V(z) \tag{5.12}$$

and

$$\frac{dV(z)}{dz} = -j\omega L_1 I(z) \tag{5.13}$$

where L_1 = inductance per unit length

C_1 = capacitance per unit length.

If we differentiate Eq.5.12 with respect to z, and substitute Eq.5.13, we get the wave equation in terms of I(z).

$$\frac{d^2I(z)}{dz^2} = -\omega^2 L_1 C_1 I(z) \tag{5.14}$$

whose solution after adding the time dependence is

$$I(z) = I_0 \exp\ j\omega(t \pm \sqrt{L_1C_1}\ z) \tag{5.15}$$

The velocity of propagation is clearly $1/\sqrt{L_1C_1}$ (5.16)

Comparing this Eq. with Eq.5.9, and recognizing that the velocity must be the same no matter how it is computed, it follows that for a transmission line wave

$$\sqrt{L_1 C_1} = \sqrt{\mu\varepsilon} \qquad (5.17)$$ m

Thus if we know μ and ε for the material inside the transmission line, we have one relation between L_1 and C_1, and we can determine the inductance per unit length from the capacitance per unit length or vice versa. Further, if we differentiate Eq.5.15 with respect to z, eliminate I_0 between Eq.5.15 and its derivative, and substitute it in Eq.5.12, we find the ratio of V to I for a propagating wave. This is the *characteristic impedance* of the line

$$Z_0 = V(z)/I(z) = \sqrt{L_1/C_1} \qquad (5.18)$$ m

To illustrate the application of these principles consider an *air filled coaxial line* with inner conductor diameter d_1 and outer diameter d_2.

Since we are dealing with a transmission line wave, the electric field vector in the transverse plane is what it would be in a static case.

$$E = E_r = q_1/2\pi\varepsilon r$$

where we have used Eq.5.4 to relate the charge per unit length, q_1, to the electric intensity. Further,

$$V = \int_{d_1/2}^{d_2/2} E_r \cdot dr = \frac{q_1}{2\pi\varepsilon} \ln (d_2/d_1)$$

The capacity per unit length is q_1/V. Therefore,

$$C_1 = 2\pi\varepsilon/\ln (d_2/d_1) \qquad (5.19)$$

By Eq.5.17,

$$L_1 = \mu\varepsilon/C_1 = \frac{\mu}{2\pi} \ln (d_2/d_1)$$

and

$$Z_0 = \sqrt{L_1/C_1} = \frac{1}{2\pi} \sqrt{\mu/\varepsilon} \ln (d_2/d_1) \qquad (5.20)$$ m

For our air filled line $\sqrt{\mu/\varepsilon_0} = 377$. Hence $Z_0 = 60 \ln (d_2/d_1)$.

5.3 Plane Waves.

A plane wave traveling in free space is, in a sense, *a transmission line mode*; E and H are transverse to the direction of propagation, the phase velocity is $1/\sqrt{\mu\varepsilon}$ and the field quantities in the transverse planes are solutions to LaPlace's equation (they are constants, independent of position). Because of the infinite extent of the wave, it is difficult to speak in terms of voltages and currents, but we do define an impedance as the ratio of E to H. To find its value, consider the Y component of the curl of H

$$(\nabla \times H)_y = \begin{vmatrix} \frac{\partial}{\partial z} & \frac{\partial}{\partial x} \\ H_z & H_x \end{vmatrix} = \frac{\partial Hx}{\partial z} \tag{5.21}$$

since $H_z = 0$ for the plane wave propagating in the z direction.

From Eq.5.9 we find

$$\frac{\partial H_x}{\partial z} = (\nabla \times H)_y = j\omega\sqrt{\mu\varepsilon}\, H_x \tag{5.22}$$

Next, substituting Eq.5.22 in the y component of Eq.5.1 with $\sigma = 0$, we have

$$E_y = \frac{j\omega\sqrt{\mu\varepsilon}}{j\omega\varepsilon} H_x = H_x \sqrt{\mu/\varepsilon}$$

Therefore

$$E_y/H_x = \sqrt{\mu/\varepsilon} = Z_0 \tag{5.23}$$

This is the ratio of E to H at the same point in the field and it has, in all respects, the characteristics of an impedance. For example, the power flow past any point in watts per square meter is given by

$$\text{Power flow} = H^2\sqrt{\mu/\varepsilon} \quad \text{or} \quad E^2\sqrt{\varepsilon/\mu} \text{ per square meter} \tag{5.24}$$

This can be shown directly from the Poynting vector and is *analogous to* I^2R *or* V^2/R in calculating the power transfer across a transverse plan in a transmission line.

5.4 Skin Depth and Conductor Losses.

Consider a plane wave traveling in a conductive medium. Equation 5.9 with $\frac{\sigma}{j\omega} + \varepsilon$ substituted for ε is

$$H_x = H_0 \exp\left[\pm j\omega\sqrt{\mu(\varepsilon+\sigma/j\omega)}\; z + j\omega t\right] \tag{5.25}$$

if $\sigma >> j\omega\varepsilon$, and dropping the time dependence for convenience, this becomes

$$H_x = H_0 \exp \pm \sqrt{j\omega\mu\sigma}\; z \tag{5.26}$$

and since $\sqrt{j} = (1+j)/\sqrt{2}$

$$H_x = H_0 \exp - \sqrt{\pi f \mu \sigma}\,(1+j)\,z \qquad (5.27)$$

where the minus sign has been selected to indicate that the wave must attenuate as it goes into the conductor. From Eq.5.27 it can be seen that the fields fall to 1/e of their initial value in a distance of

$$\delta = \frac{1}{\sqrt{\pi f \mu \sigma}} \qquad (5.28)$$ m

This is the *skin depth,* and we can write $H_x = H_0 \exp - (i+j)z/\delta$

Further, from Eq.5.23 with $\varepsilon + \sigma/j\omega$ substituted for ε and $\sigma >> j\omega\varepsilon$ we find

$$Z_0 = E_y/H_x = \sqrt{j\omega\mu/\sigma} = \frac{\sqrt{2j}}{\sigma\delta} = \frac{1+j}{\sigma\delta} \qquad (5.29)$$

The wave described by Eq.5.27 is propagating in the z direction with currents flowing in the y direction. Let z=0 be the surface. The *surface current density* J(x,y) in a surface perpendicular to the z axis is defined as the integral in depth of all the current which flows through an area which is of unit length in the x direction, and extends far enough into the conductor (in the z direction) that the currents beyond it are negligable. Thus for such an area which is perpendicular to the y direction (the current flow direction);

$$J_y = \int_0^\infty \sigma E_y\,dz = \sqrt{j\omega\mu\sigma} \int_0^\infty H_x\,dz \qquad (5.30)$$

where we have used Eq.5.29. From Eq.5.27,

$$J_y = \sqrt{j\omega\mu\sigma}\,H_0 \int_0^\infty \exp\left[-\sqrt{\pi f \mu \sigma}\,(1+j)\,z\right] dz = H_0 \qquad (5.31)$$

where the last equality can be proven by direct integration. Thus $J_y = H_0$

Therefore the *total surface current which flows is equal to the value of* m
the tangential magnetic field at the surface. Note that this equality is quite independent of the conductivity. In the limit of high conductivity, the skin depth is 0 and the current is a true surface current; if the conductivity is not infinite, the current will be spread in depth but its total value will be the same.

From Eq.5.29 we can now show that the value of E_y at the surface is related to the surface current density by

$$\frac{E_0}{J_y} = \frac{E_0}{H_0} = \sqrt{j\omega\mu/\sigma} = \frac{i+j}{\sigma\delta} \qquad (5.32)$$ m

This is called the *surface impedance.* The real part of this impedance or the resistive part, is exactly what one would expect in a *slab of thickness* δ *which has a conductivity* σ. Therefore, the in-phase part of the voltage at the surface is exactly the same as if the current were d.c. and it flowed uniformly across a slab of the conducting material with a thickness δ.

Even more significant is the fact that the *conductive losses per unit surface area are exactly what one would expect in that case.* This can be shown as follows using Eq.5.29. The total loss per unit of surface area is

$$L_a = \sigma\int_0^\infty EE^* \, dz = \sigma\int_0^\infty \frac{1+j}{\sigma\delta}\,\frac{1-j}{\sigma\delta}\, H(z)\, H^*(z)$$

where E^* = complex conjugate of E

H^* = complex conjugate of H.

Using Eq.5.27, integrating, and recognizing from Eq.5.31 that $H_o = J_y$

$$L_a = \frac{2}{\sigma\delta^2} H_o^2 \int_0^\infty \exp\,(-2z/\delta)\, dz = \frac{1}{\sigma\delta} H_o^2 = \frac{1}{\sigma\delta} J_y^2 \qquad (5.33)$$

The resistance of a slab of unit length and width and thickness δ is $1/\sigma\delta$. *Therefore the losses in a conductor much thicker than the skin depth are the same as if a d.c. current of magnitude equal to the surface current or the tangential component of H at the surface, flowed uniformly in a slab of the material of thickness δ.*

5.5 Attenuation Due to Finite Conductivity.

The relation between the losses per meter of transmission line and the attenuation of that line in dB per meter can be derived as follows. We start with

$$L_d(z) = -\frac{dP(z)}{dz} \qquad (5.34)$$

where L_d = is the loss per meter

$P(z)$ = is the power being transmitted in the z direction along the line.

However, as the power propagates along the line, the losses will not remain constant, rather they will decrease as the net power decreases due to attenuation. What we expect to remain constant is:

$$\frac{L_d(z)}{P(z)} = \gamma \qquad (5.35)$$

since the losses are porportional to the square of the currents and hence will decrease in linear proportion to the power transmitted.

Combining Eq.5.34 and 5.35 so as to eliminate $L_d(z)$ gives

$$\gamma P(z) = \frac{-dp(z)}{dz}$$

whose solution is

$$P(z) = P_o \exp - \gamma z$$

where γ is the attenuation per meter in nepers. To convert to dB, note that

$$e^{-1} = 10^{-.434} \text{ or } 4.34 \text{ dB.}$$

Therefore, the attenuation in dB per meter is given by 4.34γ. Combining this result with Eq.5.35 and Eq.5.33, the attenuation in dB per meter is

$$\alpha = \frac{4.34}{P(z)} \int \frac{J_s^2(z)}{\sigma\delta} \, dA \qquad (5.36)$$

where the integration is over all surfaces of the transmission line contained in one meter of its length.

Now let us consider applying this to our air filled coaxial line. Of course, if the conductors have a finite conductivity there must be a finite component of E in the direction of transmission, and we cannot have a true TEM or transmission line wave. However, if the losses per unit length are small, the field configuration will be essentially the same as for a lossless line and the transmission line mode approximation will be reasonably valid.

Using the treatment of the air filled coaxial line given in Section 5.2, the line current will be uniformly distributed over the inner and outer conductors. Therefore the inner conductor $J_s(z) = I(z)/\pi d_1$ and for the outer conductor it will be $I(z)/\pi d_2$. Writing I for I(z) the average power loss per meter on the inner conductor will be, by Eq.5.33:

$$L_{d_1} = \pi d_1 \, (I/\pi d_1)^2/\sigma\delta$$

$$L_{d_1} = I^2/\pi d_1 \sigma\delta$$

Since the same current flows in the outer conductor, the total conductive loss per unit length is

$$L_d = L_{d_1} + L_{d_2} = \left(\frac{1}{d_1} + \frac{1}{d_2} \right) I^2/\pi\sigma\delta \qquad (5.37)$$

Notice, however, that the resistance of a one meter long annular tube at the surface of the inner conductor of thickness δ and conductivity σ is $1/\pi\sigma d_1\delta$. For the outer conductor, a similar tube will have resistance $1/\pi\sigma d_2\delta$. Therefore the loss per unit length in the coaxial line can be expressed as

$$L_d = I^2 R \qquad (5.38)$$

where R, which is the effective r.f. resistance of the coaxial line per unit length, is actually *the resistance per unit length of a tube of thickness* δ, the skin depth. The current flows in the inner and outer conductors in series. m

To complete the treatment notice that the average power transmitted across any transverse plane, by virtue of the discussion in Section 5.2, is $V(z)I(z)$ or $I^2(z)Z_0$ so that by Eq.5.35

$$\gamma = \frac{I^2R}{I^2Z_0} \qquad (5.39)$$

In dB the attenuation is therefore

$$\alpha = 4.34\ R/Z_0\ \text{dB/meter.} \qquad (5.40)$$

where

$$R = \left(\frac{1}{\pi d_1 \sigma\delta} + \frac{1}{\pi d_2 \sigma\delta}\right)$$

5.6 Waveguide Propagation.

At the surface of a conductor the boundary conditions require that the tangential component of *E* and the normal component of *H* must go to zero. If some sort of field configuration is to propagate inside a closed conductor, it must have non-zero values somewhere in the interior for some component of both *E* and *H*; however each component must go to zero at some points on the boundary, i.e., where the boundary is either parallel to or perpendicular to the component. Since it is a closed conductor, at least two such points must exist. Thus, for example, consider a circular waveguide in which a wave propagates in the z direction. The x component of *H* must go to zero whenever the boundary is perpendicular to the x axis. Since there are 2 such points in the cross section of the guide, H_x must be zero at these two extreme points and non-zero someplace in between. This can only happen if $\partial^2H_x/\partial x^2 = -aH_x$ where 'a' is positive. Recall the discussion of the wave equation in Section 5.1. Furthermore, 'a' must have some definite value because its value will determine the cycle length for the variation of H_x in the transverse direction, and an exact multiple of half cycles must fit into the dimension of the waveguide. Thus, if we rewrite Eq.5.8 for the case of the waveguide in the form

$$\frac{\partial^2 H_x}{\partial x^2} + \frac{\partial^2 H_x}{\partial y^2} + \frac{\partial^2 H_x}{\partial z^2} = -\omega^2\mu\varepsilon H_x \qquad (5.41)$$

and we consider propagation in the z direction, the waveguide boundary conditions will set some definite values for the first two derivatives in the equation. Although one of these may be zero, in accordance with the above discussion, the sum must be negative if the field configuration is to fit the boundary conditions. Therefore, if we designate the sum of these two second derivatives as $-k^2H_x$, we have, for the variation in the direction of propagation

$$\frac{\partial^2 H_x}{\partial z^2} = -(\omega^2\mu\varepsilon - k^2)H_x \qquad (5.42)$$

This is a *wave equation* only for $\omega > k/\sqrt{\mu\varepsilon}$. For $\omega = k/\sqrt{\mu\varepsilon}$ it becomes *LaPlace's equation* whose solution is a straight line, and for $\omega < k/\sqrt{\mu\varepsilon}$ it becomes a *hyperbolic differential equation.*

The solution of this equation is,

$$H_x = H_0 \exp j\omega\left(t \pm \sqrt{\mu\varepsilon - k^2/\omega^2}\ z\right) \qquad (5.43)$$

One can readily see that the wave length is longer than c/f as given by Eq. 5.9, and further, it depends upon the value of ω. For larger values of ω the difference is small, but as ω becomes less and less, it reaches a value such that the square root in Eq.5.43 becomes 0; there is no longer any z dependence and Eq.5.42 becomes LaPlace's equation. This corresponds, of course, to the *cutoff frequency* and, in fact, Eq.5.43 can be rewritten as

$$H_x = H_0 \exp - j\omega \left(t \pm \sqrt{\mu\varepsilon}\sqrt{1 - f_c^2/f^2}\, z\right)$$

where $f_c = k/2\pi\sqrt{\mu\varepsilon}$

This equation has the advantage of clearly showing how the propagation depends upon the proximity to the cutoff frequency, and of being independent of guide and wave configurations.

The velocity of *phase propagation* is

$$V_p = \frac{1}{\sqrt{\mu\varepsilon}} \frac{1}{\sqrt{1 - f_c^2/f^2}} \tag{5.44}$$

which is dependent on frequency and hence *dispersive*, unlike the TEM waves described in 5.2. Note also that the phase velocity is higher than that of light. However, the *group velocity* which is given in general by

$$\frac{1}{v_g} = \frac{d}{df}\left(\frac{f}{v_p}\right) \tag{5.45}$$

in this case is

$$v_g = \frac{1}{\sqrt{\mu\varepsilon}} \sqrt{1-(f_c/f)^2} \tag{5.46}$$

which is always less than the velocity of light.

The TE_{10} wave is the most common waveguide mode used. The fields (both *E* and *H*) have no variation in the direction of *E* and go through 1/2 cycle in the direction which is perpendicular to *E* (and perpendicular to the direction of propagation). The *H* field has some component in the direction of propagation, but the *E* field is purely transverse. This is the reason for the designation TE which stands for *"transverse electric"*. The cutoff wave length is twice the width of the guide (the width being the direction perpendicular to the *E* vector) so that the cutoff frequency is given by

$$f_c = \frac{1}{2a\sqrt{\mu\varepsilon}} \tag{5.47}$$

The conductive losses in such a guide and the attenuation can be calculated by the method described in Section 5.5. The result is

$$\alpha = \frac{4.34}{\sigma\delta\sqrt{\frac{\mu}{\varepsilon}}\, b\sqrt{1-f_c^2/f^2}} \left[1 + \frac{2b}{a}\left(\frac{f_c}{f}\right)^2\right] \text{ dB/meter} \tag{5.48}$$

where σ = conductivity of the metal

δ = skin depth

$\sqrt{\mu/\varepsilon}$ = 377 ohms for air

b = dimension in the E direction, the short dimension of the guide cross section

a = dimension perpendicular to E, the larger cross section dimension.

5.7 Ionospheric Propagation.

The ionosphere is a region in which the molecules of air are ionized by ultraviolet radiation and soft x-rays coming from the sun, and the recombination rates of electrons and ions are low because of the low density of the gas. This results in a very significant population of charged molecules and atoms, of the order of 10^{10} to 10^{12} per cubic meter.

The peculiarities of radio and microwave propagation in the ionosphere do not, however, arise form the ions, *but from the electrons which are present.* It is the separation of these electrons from neutral atoms which causes the ions, and the positive ions and electrons are present in equal density so that the entire plasma is neutral in electrical charge. The ions, because of their great mass compared to that of the electrons, are not as easily accelerated by electric fields and hence have no significant effect upon the electromagnetic properties of the medium.

The principal peculiarity of propagation in the ionosphere which is to be derived in this section is that it is quite *opaque for frequencies below about 1 megahertz* and reasonably transparent to frequencies above 10 MHz. One might suppose that the free electrons, because they are mobile in the electric field, would cause the ionosphere to behave like an ordinary conducting or lossy medium. However, the situation is much more complex and interesting.

To derive the properties of ionospheric propagation, we start with Newton's equation of motion for an electron in an electric field and subject to a velocity dependent drag force.

$$m\frac{dV}{dt} = -eE\,(x,t) - k_d\,V \tag{5.49}$$

where m = electron mass (9.1×10^{-31} Kg)

V = its velocity

e = electron charge (1.6×10^{-19} coulombs)

$E(x,t)$ = electric field intensity (volts/meter)

k_d = damping constant (kilograms/meter/second).

If $E = E_0 \exp\,(j\omega t)$, the solution of Eq.5.49 is

$$V = \frac{e\,E_0 \exp\,(j\omega t)}{m\,(g + j\omega)} \tag{5.50}$$

where $g = k_d/m$.

If there are n electrons per cubic meter, the total current which flows is $-Ven$ or

$$I = ne \frac{e\, E_0 \exp j\omega t}{m\, (g + j\omega)}$$

$$I = \frac{g - j\omega}{m\, (g^2 + \omega^2)}\; ne^2 E \qquad (5.51)$$

The ratio of I to E can be considered as a complex conductivity

$$\sigma = \frac{g - j\omega}{m\, (g^2 + \omega^2)}\; ne^2 \qquad (5.52)$$

where the real part is in phase with the voltage like ordinary conduction currents, but the imaginary part is inductive.

We can again use the rule we adopted near the beginning of the chapter to substitute $\varepsilon + \sigma/j\omega$ for ε in any of the propagation equations to find the effect of conductivity upon any propagation characteristic. Using Eq. 5.52 for σ, the substitution which we must make is

$$\varepsilon \rightarrow \varepsilon_0 - \frac{ne^2\, (\omega + jg)}{m\omega\, (g^2 + \omega^2)} \qquad (5.53)$$

Before doing this, however, we can greatly simplify the mathematics for the present case because the damping in the ionospheric case is extremely small so that above audio frequencies g is not significant relative to ω except for a special case which we will note. For the frequency range of interest, therefore,

$$\varepsilon = \varepsilon_0 - ne^2/m\omega^2 \qquad (5.54)$$

At very high frequencies, therefore, the displacement current dominates and ε is close to ε_0, the dielectric constant of free space. At very low frequencies, however, ε changes sign; the total current is 180° out of phase with the displacement current. Someplace in between, *the total current in the medium (displacement plus conduction) goes to zero.* The frequency at which this occurs is called the *plasma frequency,* and from Eq.5.54 it can be seen to be

$$f_p = \frac{e}{2\pi}\; \sqrt{n/m\varepsilon_0} \qquad (5.55)$$

We can substitute Eq.5.55 in Eq.5.54 to find

$$\varepsilon \rightarrow \varepsilon_0\; [1 - (f_p/f)^2] \qquad (5.56)$$

which is equivalent to Eq.5.53 and can be used in any of the other relationships derived from Maxwell's equations to find the ionospheric effect. For example for a plane wave the phase velocity of propagation is $1/\sqrt{\mu\varepsilon}$ (see Eq. 5.9). If we make the substitution indicated by Eq.5.56 the *phase velocity* in the ionosphere is given by

$$V_p = \frac{1}{\sqrt{\mu\varepsilon}} \rightarrow \frac{1}{\sqrt{\mu\varepsilon_0}} \quad \frac{1}{\sqrt{1 - f_p^2/f^2}} \tag{5.57}$$ m

Compare this with Eq.5.44. The only difference is that the *plasma frequency replaces the cutoff frequency of the waveguide.* The similarity is, of course, not physical. The phenomena are vastly different, but the relation ship is revealing and useful. At frequencies below the plasma frequency, signals will not propagate, just as they will not propagate in a waveguide below cutoff. Signals of frequency lower than the plasma frequency will therefore be totally reflected from the ionosphere.

The *group velocity* in the ionosphere is given by Eq.5.46 with f_p replacing f_c. It approaches zero as the frequency approaches the plasma frequency.

The *wave impedance* is, from Eq.5.23 and 5.56

$$Z_0 = \frac{E}{H} = \sqrt{\frac{\mu}{\varepsilon}} \rightarrow \frac{\sqrt{\mu}}{\sqrt{\varepsilon_0}[1-(f_p/f)^2]} \tag{5.58}$$

The significance of Eq.5.58 is, of course, that the wave impedance approaches infinity as the frequency approaches the plasma frequency. This is to be expected from our previous discussion since the total current, displacement plus conduction, is zero. This is like a parallel resonance phenomenon. The current carried by the electrons lags the voltage and hence is inductive. The ordinary displacement current is, of course, capacitive. One would expect that at the plasma frequency the small amount of damping which we have previously neglected would be significant. In fact it is; however, in the ionosphere the damping is sufficiently small that the wave impedance at the plasma frequency and phase velocity are very large indeed and the group velocity is quite small.

The reflection coefficient at the interface between two media, or transmission lines of characteristic impedance Z_1, and Z_0 is given by

$$\Gamma = \frac{Z_1 - Z_0}{Z_1 + Z_0} \tag{5.59}$$ m

This equation is not always directly applicable to ionospheric reflections because the ionosphere does not begin abruptly. Reflections may be small if the transition in electrical characteristics is gradual. However, below the plasma frequencies the reflections are almost total. It is this phenomenon that provided the means by which the existance of the ionosphere was first experimentally demonstrated, and the demonstration by *Breit and Tuve* in 1926 was the first application of radar (Ref.22). Working at radio frequencies, *they transmitted pulses of radiation upward to the ionosphere, and received echoes.* They calculated the height of the ionosphere from the time of travel. The echo return was quite large because of the large physical extent of the ionosphere, and the experiment could be performed with the rudimentary equipment of that time. It is interesting to note that this event took place before the basic principles of thermal noise had been discovered. (See the dates of References 1 and 2.)

Since the density of electrons in various ionospheric layers normally varies between 10^{10} and 10^{12}, the plasma frequency (from Eq.5.55) will vary between

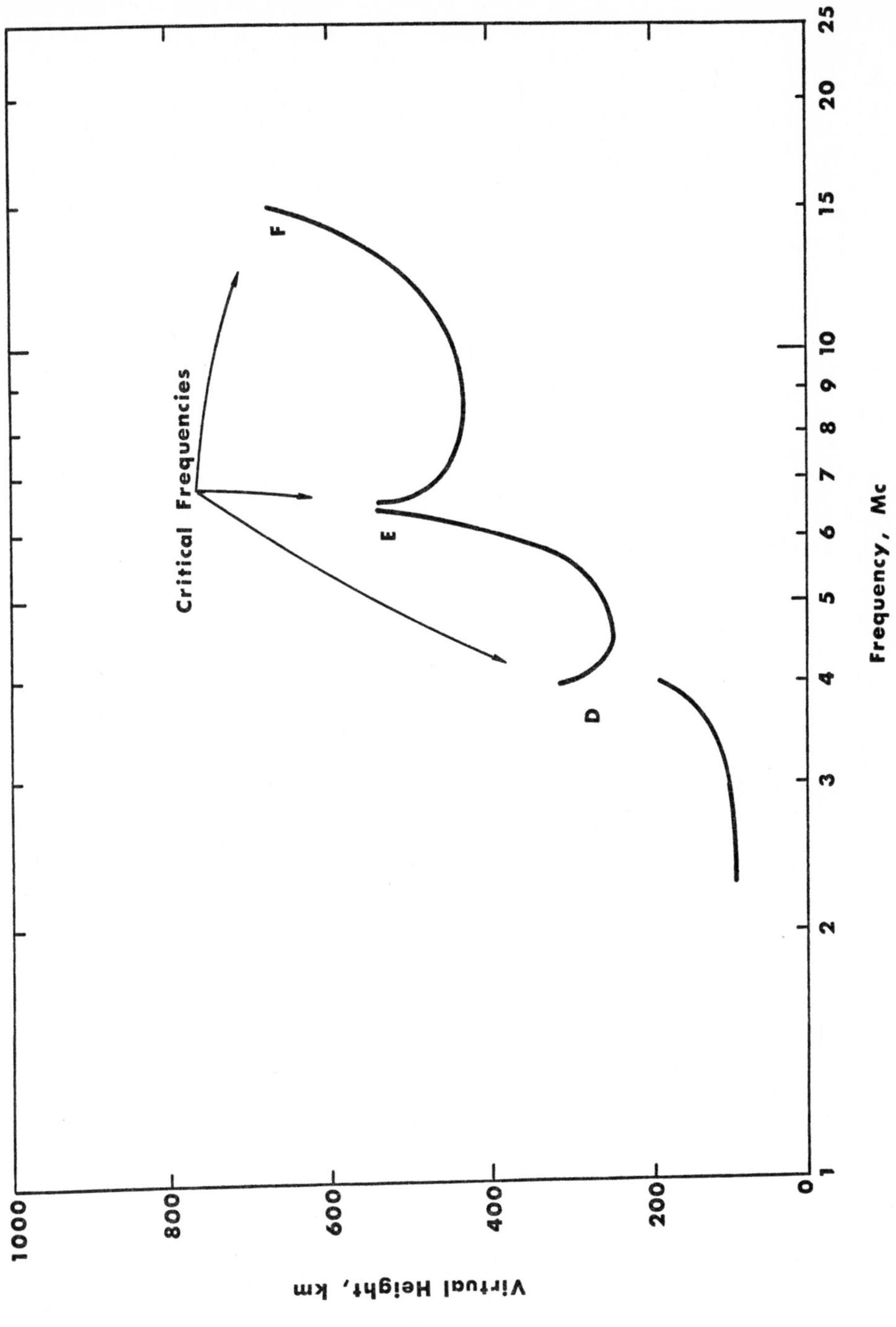

Fig. 5.1. Ionospheric Sounder Output

$$\frac{1.6 \times 10^{-19}}{2\pi} \sqrt{\frac{10^{10}}{9.1 \times 10^{-31} \times 10^{-9}/36\pi}} = 0.9 \text{ MHz}$$

and 9 megahertz.

There are commercially available devices called *ionospheric sounders*, which provide data on the electron density in the ionosphere as a function of height. These are essentially upward looking pulse radars which vary pulse frequency slowly so that in about 1 minute they cover the frequency range from 1 to 25MHz. They provide, as data output, a plot of echo delay (time from pulse to echo, which is a measure of the distance to the reflecting layer) as a function of pulse frequency. A typical plot of this type is shown in Fig.5.1. The ordinate is labeled virtual height because it is somewhat larger than true height due to the fact that the group velocity is lower than the free space velocity. The discrepancy is particularly large in the vicinity of the plasma frequency, which gives rise to the cusp shapes in Fig.5.1. Also, the echo amplitudes are greatly reduced due to conductive losses in the immediate vicinity of the plasma frequency. This accounts for the missing portions of the curve.

The D, E, and F layers are in that order of height and also in order of increasing densities of electrons. This permits the ionospheric sounder to locate them. It obviously cannot locate a layer with low f_p above a high f_p region. There are, however, other techniques which can do this, although they are not as simple and accurate. In daytime the ionization is generally greater than at night, particularly for the lower layers where the rate of recombination of ions is rapid. The echo sounder curve will move to the right at night, but it will also change substantially in shape. The reason for the detailed and distinct layer structure relates to the chemistry of the atmosphere, ionization and recombination rates, and the radiation spectrum of the sun.

Obviously this treatment of ionospheric propagation covers only a single elementary aspect. We have ignored, for example, the effect of the earth's magnetic field on the electron motion, oblique propagation and focusing effects, scatter propagation and absorption effects. A comprehensive treatment of these subjects plus details of the formation of the ionized layers is given in Ref.25.

5.8 Extraterrestrial Propagation.

The space around the earth and even interstellar space is not truly free space because it is not a perfect vacuum. Although interstellar space is a better vacuum than any we can make here on earth, it still contains atoms which are ionized by the radiation present; and because of the low density of matter, the recombination rates are very low so that *ions and electrons are present*. As in the ionosphere, the electrons, because of their high mobility, affect electromagnetic propagation much more than the ions and in the same way. The electron densities in interstellar space are believed to be of the order of 10^5 per cubic meter (Ref.23). In the cislunar space (between the earth and the moon) the electron density is believed to be of the order of 10^9 per cubic meter.

The electron density in the cislunar space has been determined by measuring the frequency dispersion or the dependence of group velocity on frequency using radar signals bounced off the moon. Thus for an electron density of $10^9/m^3$, the plasma frequency will be, from Eq.5.55

$$f_p = \frac{e}{2\pi} \sqrt{n/m\varepsilon_o} = .282 \text{ MHz}$$

The group velocity (from Eq.5.46 with f_p replacing f_c) is given by

$$v_g = \frac{1}{\sqrt{\mu\varepsilon}} \sqrt{1 - (f_p/f)^2} \tag{5.60}$$

which, when $f_p << f$ becomes

$$V_g = c\,[1 - \frac{1}{2}(f_p/f)^2] \tag{5.61}$$

where c = velocity of light in free space

f_p = plasma frequency

f = signal frequency.

At 20 MHz, in the cislunar space, $1/2\,(f_p/f)^2 = 10^{-4}$ Therefore the plasma effect will cause a 20 MHz signal to travel at a velocity less than that of free space by .01%.

The distance to the moon is 3.84 x 10^5 Km and the time required for the round trip at the velocity c is 2.56 seconds. The additional delay due to plasma effects is therefore .25 millisecond which is easily measurable by comparing the travel time against that of a higher frequency signal (e.g., 200 MHz) for which the plasma effect is very much smaller.

The effect of interstellar electrons is most clearly observed in the signal arrivals from *pulsars* which are stars which put out radio signals at extremely regular intervals which vary from pulsar to pulsar, but are of the order of 1 second. The received *signals are FM slides,* starting at high frequencies and becoming more low pitched with time.

Astronomers believe these are very dense, rapidly rotating neutron stars and the signals start out as a simple step function or spike, appearing as FM slides only because the *dispersion* in the interstellar space path seperates out the frequency components according to their group velocities. Since the density of electrons in interstellar space is about $10^5/m^3$, the plasma frequency will be about 2.8KHz, and the group velocity at 40 MHz will be less than c by a factor of $(1 - 2.45 \times 10^{-9})$. This is a very small reduction in velocity, but for a pulsar at the modest distance of 400 light years away, the delay is of the order of $400 \times 2.45 \times 10^{-9} = 10^{-6}$ years or 31 seconds. Thus we would expect to find that the 40 MHz portion of the signal arrives about 31 seconds after the very high frequency portion. In fact, it does, and the entire shape of the pulse is such that the delay is inversely as the square of the frequency exactly as one would expect from Eq.5.61, thus confirming that plasma dispersion is the origin of the shape of the signal (Ref. 24).

CHAPTER VI

REFLECTION REFRACTION AND SCATTERING

6.1 Interfaces Between Homogeneous Media#

The electromagnetic boundary conditions at an interface between two media require that the tangential component of E and the normal component of H be continuous through the surface. This must be true at each point of the surface, which requires, first of all, that the *wave length measured along the surface be the same for incident, reflected, and refracted waves.* By the wave length measured along the surface, we mean the distance that would be measured along the surface from peak to peak or from trough to trough at any particular instant of time. As can be readily seen from Fig.6.1, this means that

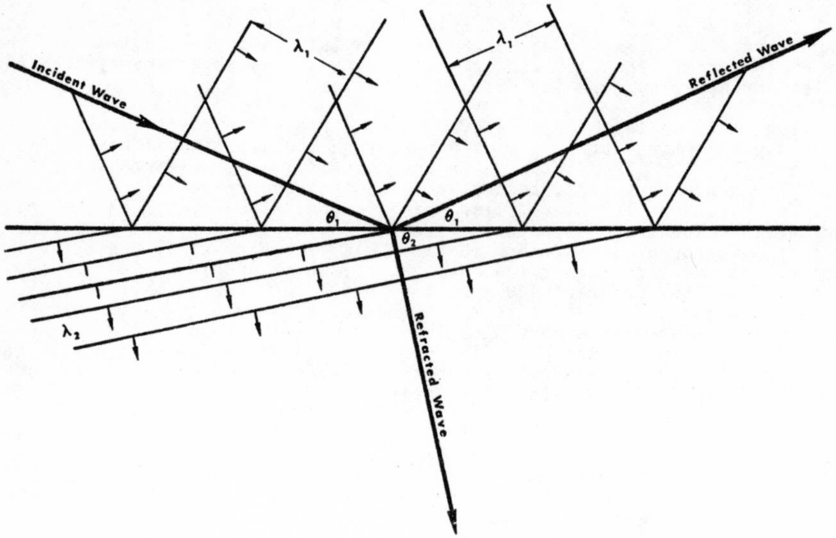

Fig. 6.1. Wave fronts and directions of propagation.

$$\lambda_1/\cos\theta_1 = \lambda_2/\cos\theta_2 \qquad (6.1)$$

and since $\lambda_1 = V_{p1}/f$; $\lambda_2 = V_{p2}/f$ where V_{p1} and V_{p2} are phase velocities,

#For a more complete discussion of this subject, see Ref.26.

$$V_{p1}/\cos\theta_1 = V_{p2}/\cos\theta_2 \tag{6.2}$$

and since $V_p = 1/\sqrt{\mu\varepsilon}$ (See Section 5.1 and 5.3)

$$\sqrt{\mu_1\varepsilon_1}\cos\theta_1 = \sqrt{\mu_2\varepsilon_2}\cos\theta_2 \tag{6.3}$$

This is *Snell's Law.* θ_1 and θ_2 are the *grazing angles,* which are the complements to the *angle of incidence and refraction.* Snell's Law is often seen with the cosines replaced by the sines of their complements.

Notice that if V_{p2} is larger than V_{p1} as it would be in transiting from a denser to a less dense medium and θ_1 is small, there may be no angle θ_2 which satisfies Eq.6.3. This case is called *total internal reflection.*

E and *H* are, of course, perpendicular to the direction of propagation. If the polarization is such that *E* is in the plane of incidence (this is the plane containing the direction of the incident wave and the normal to the surface) the requirement that the tangential component of *E* be continuous implies that

$$E_i\sin\theta_1 + E_r\sin\theta_1 = E_t\sin\theta_2 \tag{6.4}$$

where the subscripts i, r, and t refer respectively to the incident, reflected and transmitted waves. (The angle of reflection is, of course, equal to the angle of incidence.)

If, on the other hand, *E* is perpendicular to the plane of incidence it will be parallel to the interface between the media. This will be true of *E* in all three waves and *H* will be in the plane of incidence. In that case the requirement on *E* is that

$$E_i + E_r = E_t \tag{6.5}$$

The corresponding requirements on *H* are given by Eq.6.6 and 6.7. For *E* in the plane of incidence

$$H_i + H_r = H_t \tag{6.6}$$

and for *H* in the plane of incidence

$$H_i\sin\theta_1 + H_r\sin\theta_1 = H_t\sin\theta_2 \tag{6.7}$$

For *E* in the plane of incidence we can write Eq.6.6 in terms of the *E* vector magnitudes and the characteristic impedance using Eq.5.23. Thus,

$$\frac{E_i}{Z_{01}} - \frac{E_r}{Z_{01}} = \frac{E_t}{Z_{02}} \tag{6.8}$$

where we recognize that $E_r = -Z_{01}H_r$ because the reflected wave is traveling in the reverse direction.

If we eliminate E_t between Eq.6.8 and 6.4, and solve for E_r we find the reflection coefficient, *for E in the plane of incidence.*

$$\Gamma = \frac{E_r}{E_i} = \frac{Z_{02} \sin \theta_2 - Z_{01} \sin \theta_1}{Z_{02} \sin \theta_2 + Z_{01} \sin \theta_1} \tag{6.9}$$

This is the familiar expression for the reflection coefficient between two transmission lines of different characteristic impedance (see for example Eq. 5.59) except for the presence of the sines of the angles. These sines can be readily calculated from Snell's Law (Eq.6.3).

If instead of eliminating E_t between Eq.6.4 and 6.8 we eliminate E_r and calculate the ratio between E_t and E_i, we find

$$T^1 = \frac{E_t \sin \theta_2}{E_i \sin \theta_1} = \frac{2Z_{02} \sin \theta_2}{Z_{01} \sin \theta_1 + Z_{02} \sin \theta_2} \tag{6.10}$$

Except for the sines of the angles, this is the familiar expression for the transmission through an interface between transmission lines of different impedances. Equations 6.9 and 6.10 can be interpreted very simply as the familiar expressions for reflection and transmission coefficients except that the *characteristic impedances are multiplied by the sine of the angle between the direction of propagation and the interface, and the coefficients apply to the component of transmission perpendicular to the interface.*

The corresponding expressions for that polarization in which the *E* vector is perpendicular to the plane of incidence can be derived from Eq.6.5 and 6.7. They are

$$\frac{E_r}{E_i} = \frac{Z_{02}/\sin \theta_2 - Z_{01}/\sin \theta_1}{Z_{02}/\sin \theta_2 + Z_{01}/\sin \theta_1} \tag{6.11}$$

for *H* in the plane of incidence and

$$T^1 = \frac{E_t}{E_i} = \frac{2Z_{02}/\sin \theta_2}{Z_{01}/\sin \theta_1 + Z_{02}/\sin \theta_2} \tag{6.12}$$

Notice in this case the *E* vectors are parallel to the interface and therefore on the left hand side of Eq.6.12 they are not multiplied by the sine of the angle as they are in Eq.6.10.

In any case, if the proper characteristic impedances are used, i.e., the *wave impedance of the medium multiplied by or divided by the sine of the angle;* transmission and reflection coefficients can be derived in the usual way, and *these coefficients apply as ratios of the components of the E vector parallel to the interface.* In this way these oblique incident waves can be treated exactly the same as transmission lines and, for example, the familiar *Smith Chart* can be used to advantage, particularly when multiple interfaces are present. First, of course, one must apply Snell's Law to find the proper angles.

Most materials which are likely to reflect microwaves in practical applications, e.g., the earth, sea, and structures of various types, have dielectric constants which differ from that of free space, but do not have permeabilities which are significantly different from that of free space. A material which has a dielectric constant larger than that of free space will have a lower characteristic impedance and a lower velocity of phase propagation. From Snell's Law, this lower velocity of phase propagation implies that at refraction at an interface from air into such a material (from Eq.6.2) the cosine of the grazing angle will decrease. Therefore the sine of the angle will increase. For such a situation, since $\sin\theta_2 > \sin\theta_1$ and $Z_{02} < Z_{01}$, it is possible for the numerator in Eq.6.9 to become zero at some particular angle of incidence so that there is no reflected wave. This angle is known as the *Brewster Angle*. The condition as seen from Eq.6.9 is

$$Z_{02}\sin\theta_2 = Z_{01}\sin\theta_1 \tag{6.13}$$

Snell's Law, of course, must also be satisfied. It can be demonstrated by algebra and trigonometry that when Eq.6.13 and Eq.6.2 are simultaneously satisfied, the *direction of reflection* (if there were to be a reflected wave) *is perpendicular to the direction of transmission of the refracted wave,* i.e., the directions of the reflected and refracted waves are perpendicular to each other. (The reflected wave, of course, has zero amplitude.)

For distilled water ($\varepsilon = 81\varepsilon_0$) the Brewster angle is 6.3° so that an incident wave with its electric vector vertical and a grazing angle of 6.3° would be completely absorbed in the water. In a real body of water however, there is always significant conductivity resulting in what is called a "false Brewster angle". We can handle the conductivity in our usual way starting with any of the preceeding equations.

If it were the permeability μ which generally varied among different materials instead of ε, there would be a Brewster Angle for the alternative polarization, i.e., for the polarization with E perpendicular to the plane of incidence.

6.2 Refraction in Non-Homogeneous Media.

If electromagnetic waves propagate in a medium which is non-homogeneous in properties, in particular where there is a variation of index of refraction with height (z), as there is in the atomsphere, one can find the path of a ray by applying Snell's Law (Eq.6.2 or 6.3). We consider the medium as being made up of a large number of very thin layers with the velocity of propagation varying slightly from one layer to the next. Thus the condition which must be satisfied by the direction of propagation of any particular wave, which we may call a *ray path* is

$$V_p(z)/\cos\theta(z) = C_h \tag{6.14}$$

where the constant C_h obviously represents the phase velocity of the medium at that height at which $\cos\theta(z) = 1$ and the particular ray is horizontal.

For a first order approximation which is usually quite adequate for atmospheric propagation, we can consider that the *phase velocity increases linearly with height.* Thus

$$V_p(z) = K + K_1 z \tag{6.15}$$

where K and K_1 are constants over some limited range of height z. Therefore from Eq.6.14 and 6.15

$$\cos\theta(z) = \frac{V_p(z)}{C_h} = \frac{K + K_1 z}{C_h}$$

For the path of the ray

$$\frac{dz}{dx} = \tan\theta = \frac{\sqrt{1-\cos^2\theta}}{\cos\theta} = \frac{\sqrt{C_h{}^2-(K+K_1z)^2}}{K+K_1 z}$$

$$dx = \frac{(K+K_1z)\,dz}{\sqrt{C_h{}^2-(K+K_1z)^2}}$$

which integrates readily to

$$\sqrt{C_h{}^2-(K+K_1z)^2}/K_1 = x + K_2$$

where K_2 is a constant of integration. Squaring both sides gives

$$C_h{}^2 = (K_1x + K_1K_2)^2 + (K + K_1z)^2$$

or

$$(C_h/K_1)^2 = (x+K_2)^2 + (z + K/K_1)^2$$

which is the *equation of a circle* of radius C_h/K_1 with center at $z=-K/K_1$ and $x=-K_2$. K_2 merely defines the point at which the ray is horizontal. If we set $K_2=0$, the ray is horizontal at x=0.

Further, from Eq.6.15, K is the velocity of propagation at the height from which we measure z and from Eq.6.14 C_h is the velocity at the height at which the ray is horizontal. If we measure z from the point at which the ray is horizontal, $K=C_h$ and the path of the ray is given by the circle#

$$(C_h/K_1)^2 = x^2 + (z + C_h/K_1)^2 \tag{6.16}$$

where C_h = phase velocity of propagation at that height where the ray is horizontal (the maximum height)

#Reference 27 contains a more detailed discussion and derivation.

K_1 = rate of increase of phase velocity with increasing height.

In ordinary atmospheric propagation, the velocity differs from that of free space by less than one part in a thousand. Therefore, the rate of change of *index of refraction* with height, which is exactly defined by

$$\frac{dn}{dz} = \frac{d}{dz}\frac{c}{V_p} = -c\frac{dV_p}{dz} / V_p^2$$

can be written to close approximation as

$$\frac{dn}{dz} \cong -\frac{1}{c}\frac{dV_p}{dz} \cong -K_1/c \cong -K_1/C_h \qquad (6.17)$$

where n = index of refraction, a function of height

z = height (in meters)

K_1 = rate of increase of phase velocity with height (meters/second/meter)

C_h = phase velocity at that point of its circular path where ray is horizontal. (meters/second)

Therefore from Eq.6.16 and 6.17 the *radius of the circular path of the ray is simply* $(dn/dz)^{-1}$ with the curvature being downward.

In terms of the important measurable characteristics of the atmosphere the *index of refraction* at microwave frequencies is given by (Ref.28)

$$n = 1 + \frac{77.6 \times 10^{-6}}{T}\left(P + \frac{4810p}{T}\right) \qquad (6.18)$$

where T = temperature in degrees Kelvin

P = pressure in millibars (thousandths of an atmosphere)

p = partial pressure of the water vapor.

The dependence of n on height for microwaves varies with location and weather, but under average conditions a first order approximation is

$$n(z) = 1 + 313 \times 10^{-6} \exp(-.044z) \qquad (6.19)$$

with z in thousands of feet. (Ref.28)

This gives a *radius of curvature at sea level* of 73 x 10^6 feet or about 3.5 times the radius of the earth. Since the curvature is downward, a microwave path which is tangent to the earth at some point rises above the earth at a slower rate than a straight line. It can be shown from geometrical consideration that the ray with radius of curvature $r_1 > r_a$ (where r_a is the radius of the earth) rises above the earth at about the same rate as though the ray were straight and the earth's curvature were given by

$$1/r_a^1 = \frac{1}{r_a} - \frac{1}{r_1} \qquad (6.20)$$

For $r_1 = 3.5\ r_a$, Eq.6.20 gives $r_a^1 = 1.4\ r_a$. This is close to 4/3 and gives rise to *the 4/3 earths radius correction rule* (Ref.29). This rule states that for average, or standard atmospheric conditions, *the height of a nearly horizontal microwave beam or ray above the earth can be closely approximated by assuming that the ray is straight and the earth has 4/3 its actual radius of curvature.*

The height (**z**) of the straight ray above a surface with downward radius of curvature r, or of a ray with upward radius of curvature r above a flat surface is approximately

$$\mathbf{z} \cong d^2/2r \qquad (6.21)$$

where d is the horizontal distance from point of tangency and **z**, d, and r are in the same units.

In a practical case where one is establishing a point to point microwave link for communication purposes, a very careful drawing is made of the terrain between the two points including all obstructions along the path. The earth is drawn as flat, and the microwave path is drawn as curved upward. It is drawn as a circle between the two end points with various radii of curvature which may exist due to atmospheric conditions in the area, to make sure there is no interference. The path is constrained, of course, by the end points rather than by the directivity of the transmitting or receiving antennas because this directivity is certain to be very broad compared to the small angular differences which will cause interference. In practical work a structure is considered to interfere with the propagation path if it comes within about 100 feet of it. However, over the usual path lengths which are apt to be 20 to 30 miles for terrestrial microwave links, it is frequently difficult to be certain whether some obstructing object is within 100 feet of the path or not. Consequently, larger distances are used as a safety factor (Ref.30).

6.3 Useful Rules for Calculating Microwave Reflections and Scattering.

We generally characterize the scattering properties of *small objects* (small relative to the propagation path lengths of interest although not necessarily small in terms of wave length) by a *scattering cross section* σ. The microwave energy that proceeds from one antenna to another over a path involving scattering from such an object is given by

$$P_3 = \frac{P_1\,G_1}{4\pi r_{12}^2}\ \sigma\ \frac{A_{e3}}{4\pi r_{23}^2} \qquad (6.22)$$

where P_3 = power at the receiving antenna (watts)

P_1 = power transmitted by the transmitting antenna (watts)

G_1 = gain of the transmitting antenna

r_{12} = distance between the transmitting antenna and the scatterer (meters)

σ = scattering cross section (square meters)

A_{e3} = effective area of the receiving antenna (square meters)

r_{23} = distance between the scatterer and the receiving antenna (meters)

Equation 6.22 is arranged so that one can readily see the spreading loss from the transmitting antenna to the scatterer, the scattering cross section and the spreading loss from the scatterer to the receiving antenna. There are, however, in Eq.6.22, three terms which depend upon the direction involved. These are G_1 and A_{e3} which were discussed in Chapter 4; and σ, which depends upon the incident and reflected directions relative to the geometry of the scattering object. To completely measure and specify σ as a function of both angles is very complicated since it is, in fact, a 4 dimensional distribution. That is, for an object which does not have symmetry, *σ at any frequency is a function of four variables;* the polar and azimuthal angles of the direction of arrival of the incoming energy and the polar and azimuthal angles of the direction of the outgoing or scattered energy. Generally we specify σ only as an average over all combinations of directions and then specify its statistical properties, i.e., by about how much it varies as the directional angles are changed. This is the case, for example, for airplanes where one is particularly interested in the back scattering properties for radar applications. Reference 30 presents a useful collection of data on scattering cross sections.

For an object with dimensions of the order of λ or greater, made of meterial which is a good reflector at the frequencies of interest (e.g., metals) a useful rule is that the *scattering cross section will be roughly equal to the projected area* of the object in the direction toward the transmitting antenna or toward the receiving antenna or someplace in between. If these cross sections differ greatly, the estimate will be very rough.

If the object is made of material which is not a perfect reflector, one should multiply this projected area by the square of the reflection coefficient at normal incidence (use Eq.6.9 with $\theta = 90°$). This will result in a value of σ which is usually slightly too low. It can be taken as a lower bound, with the rough upper bound being the cross section estimate for a good reflector of the same size and shape.

In some cases, however, objects of the order of 1 wavelength in size can resonate, giving scattering cross sections several times larger than would be indicated by the previous paragraph. For example, a conducting sphere of radius $a = \lambda/2\pi$ has a scattering cross section of about $4\pi a^2$. Other resonances and antiresonances occur for larger spheres but they are less significant and as 'a' becomes larger relative to λ, the cross section deviates less and less from πa^2.

Objects much smaller than one wavelength in dimension are called Rayleigh Scatterers (named after the same Lord Rayleigh as the Rayleigh distributions in section 2.4). If the material of the scatterer is a good reflector, the cross section is given approximately by

$$\sigma = 10^4 \left[\frac{a}{\lambda}\right]^4 \pi a^2 \qquad (6.23)$$

where a is a significant dimension, i.e. the radius for a sphere.

The inverse dependence on the fourth power of the wavelength indicates that such scatterers become rapidly less effective as the wavelength becomes longer. It was the theoretical discovery of this relationship for light scattered by small dust particles which permitted Lord Rayleigh to explain the blue of the sky and the redness of sunsets. Equation 6.23 is quite accurate for spheres if $\lambda/a > 10$.

If the material of which the object is made is not a good reflector, the same correction (Γ^2 at normal incidence from Eq.6.9) as is made for the larger reflectors will give a good estimate.

To find the specular reflection from a flat surface which is essentially infinite in extent, or contains many Fresnel zones, the *image method* is most useful. The criterion for use of the image method is as follows: the specular path (angle of incidence equals angle of reflection) must intersect the flat surface and, if one draws a straight line from the source antenna to some point on the plane reflecting surface and thence to the receiving antenna and calculates the total line length, this length must change by several wavelengths as the point moves around on the plane surface. Further, the surface must have significant extent in both dimensions - height and width. (The ground below an elevated antenna usually satisfies this condition.) In this case the power received by reflection from this plane is given by

$$P_3 = \frac{P_1 G_1 A_{e3}}{4\pi r_i^2} \Gamma^2 \qquad (6.24)$$

where r_i = distance from the receiving antenna *to the image* of the transmitting antenna; $r_i = r_{12} + r_{23}$ where the distances are measured from each antenna to the specular reflection point.

Γ = reflection coefficient of the material of the surface at normal incidence (1 if the material is a good conductor).

A third case of particular interest is a *flat area* so oriented as to reflect energy specularly from the transmitting to the receiving antenna, and being of dimension large compared to a wavelength but *in a single Fresnel zone*. The single Fresnel zone criterion is roughly as follows: If one draws a line from the source antenna to some point on the reflecting surface, and from there to the receiving antenna, and moves the point all over the surface, one should find that the total line length changes by less than 1/4 wavelength.# If this single Fresnel zone criterion is met, the reflection from the surface can be approximated by Eq.6.22 with

$$\sigma = \frac{4\pi A_s^2}{\lambda^2} \qquad (6.25)$$

where A_s = area of the surface

Notice that σ is now proportional to the square of the area (Ref.31).

6.4 Water In the Air.

The water molecule is not symmetric. From the classical point of view, the two hydrogen atoms are not on opposite sides of the oxygen atom. The lines connecting the oxygen and hydrogen atoms are 120° apart rather than 180° apart. Because of this assymmetry the molecule is an electric dipole which seeks to orient itself in an electric field. This gives rise to a large dielectric constant ($\varepsilon = 81\varepsilon_0$) at low frequencies, and large dielectric losses at high microwave frequencies because of the friction associated with the rotation. As the frequency increases further, the rotation of the water molecules is greatly impeded due to their rotational inertia, just as the linear motion of the electrons in the ionosphere is impeded by their mass at adequately high frequencies, and water again becomes transparent. However, for water this does not occur until we reach the frequency of visible light. From the quantum mechanical point of view the phenomenon is described by rotational resonances in the water molecule in the vicinity of 22 GHz which are fairly sharp when the water is in the form of a gas, but when it is in liquid form, they are sufficiently broad that there is significant absorbtion down to 4 GHz.

#A rough test for the Fresnel zone criterion is whether or not $\pi\lambda d'/4A_s > .4$, where d' is the distance from the reflector to the transmitter or receiver, whichever is closer. If the quantity is greater than 0.4, the reflector is in a single Fresnel zone and Eq.6.25 is applicable. If not, σ will be less. It is also necessary, of course, that the surface contain the specular reflection point.

This behavior of the water molecule and the characteristics of Rayleigh scattering noted in section 6.3 explain most of the important effects of atmospheric water on microwave systems. Thus, water vapor in the atmosphere causes no scattering and very little absorbtion. At frequencies below 10 GHz, the absorbtion of the atmosphere is less than .03 dB per mile and most of that is due to oxygen. At higher frequencies there is a gradual rise in attenuation to about 1 dB per mile at 22 GHz, if the humidity is high. The principle effect is refraction as indicated by Eq. 6.18.

When water droplets are present, however, the situation is much more severe. Since water is a good reflector at all microwave frequencies (Γ is large at low frequencies because of the high dielectric constant, and at higher frequencies because of the losses), the scattering is primarily dependent on droplet size in accordance with Eq.6.23. In a very heavy rain, the total scattering cross section (the scattering cross section per drop times the number of drops) at 10 GHz may be as much as .01 Cm^2 per meter of path. In a light rain it will be about 10^{-5} cm^2 per meter of path. Fog or cloud will cause even less scattering. This amount of scattering is quite significant to radar applications and even light rain or heavy fog can be seen in 3 cm radars, particularly if they are designed to detect it, as are weather radars. In 10 GHz radars designed for other purposes, the water shows up as "weather clutter" and interferes with echo detection. At 3 GHz, as indicated by the fourth power law of Eq.6.23, the effect is about two orders of magnitude less. However, 10 cm radars are still capable of rain detection and some are designed for that purpose.

The corresponding attenuations are higher than one would expect if scattering were the only mechanism for loss of energy. At 10 GHz the attenuation in a heavy rain may be as much as 1 dB per mile and in a light drizzle .001 dB/mile. The corresponding figures at 3 GHz are again about two orders of magnitude less.#

The numbers given above are for sea level or close to it. In considering the total weather effect on long range microwave paths it is important to take account of the path elevation. Rain and other weather effects rarely extend above the height of 20,000 feet.

#Further details of microwave attenuation and scattering by rain and fog are given by Barton and Skolnick in their books cited as Ref.16 and 28.

CHAPTER VII

SOME SYSTEM CHARACTERISTICS

7.1 Modulation, Bandwidth and Noise.

The equation for the instantaneous voltage of an *amplitude modulated* signal can be written

$$v = V_0 (1 + m \sin 2\pi f_m t) \sin 2\pi f_c t \qquad (7.1)$$

where m = fractional modulation

f_m = modulation frequency

f_c = carrier frequency.

By expanding the product of the two sines by trigonometric identities we obtain

$$v = V_0 \sin 2\pi f_c t + \frac{m V_0}{2} \cos 2\pi(f_c - f_m) t - \frac{m V_0}{2} \cos 2\pi(f_c + f_m)t \qquad (7.2)$$

Thus, if the modulation is a single tone, the modulated signal will consist of three tones, carrier and upper and lower sidebands. If the modulation is a band of tones from a very low frequency up to some maximum frequency f_M, the modulated signal will contain frequencies from $f_c - f_M$ to $f_c + f_M$. It will have twice the bandwidth of the original signal. The transmission of ordinary amplitude modulated signals is wasteful in terms of signal power and bandwidth because all the information in the modulation is contained in a *single set of sidebands*, upper or lower. Several techniques are available for suppressing the carrier and one set of sidebands; e.g., a balanced modulator such as that shown in Fig.7.1 will suppress the carrier and a sharp skirted filter can be used to remove the upper or lower set of sidebands. On reception, the carrier is added before the signal is detected, and if the added carrier is within a few cycles of the original, and adequately large compared to the received signal, the detected signal will be identical to the original modulation because the detection process, which is non-linear, will then produce as its dominant terms

$$As + Bs^2 + \ldots$$

where s is the signal impressed on the detector. Assuming that this is the upper sideband plus the locally generated carrier frequency,

$$s = a \sin 2\pi f_c t + b \cos 2\pi(f_c + f_m)t.$$

The linear "A" term in the detector output simply reproduces s which has no audio or video components. The square or "B" term produces

$$B\,a^2[\text{Sin } 2\pi f_c t]^2 + B\,b^2[\text{Cos } 2\pi(f_c + f_m)t]^2$$

$$+ \quad 2B\,ab\,\text{Sin}(2\pi f_c t)\ \text{Cos } 2\pi(f_c + f_m)t.$$

Of these three terms, the first two can be shown by trigonometric identity to have d.c. and double r.f. frequency components, but no video or audio component. The third term, being the product of two sinusoids of different frequency can be decomposed into sum and difference frequencies. It is only the difference frequency which has an audio or video component. This is

$$B\,ab\ \text{Sin } 2\pi f_m t$$

which, except for the multipliers, is the original modulation.

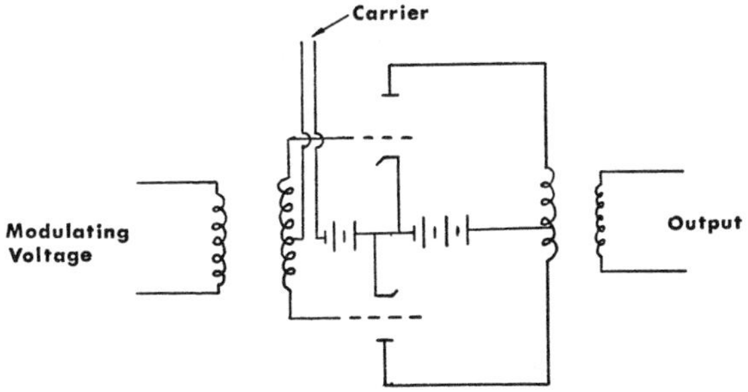

Fig. 7.1 -- Balanced modulator for carrier suppression.

Single side band amplitude modulation is widely used in modern communication systems because of its efficient use of channel space and transmitter power. The carrier frequency is sometimes transmitted at reduced power as a synchronizing signal for the regeneration of the carrier in the receiver. When many single sideband signals are multiplexed together the carriers are usually maintained in some known and reproducable relation to each other so that the receiver can regenerate all the necessary carriers from one or two synchronizing signals. Single sideband demodulation does not affect the signal to noise ratio. That is, the signal to noise ratio in the demodulated signal is the same as that in the r.f. band whose bandwidth is that of the modulation intelligence.

The equation for the instantaneous voltage in a *frequency modulated* signal is given by

$$v = V_0 \sin\,[2\pi f_c\ t + \frac{\Delta F}{f_m} \sin\,2\pi f_m\ t] \qquad (7.3)$$

where ΔF = peak frequency deviation.

Notice that the time derivative of the quantity in the square brackets is $2\pi[f_c + \Delta F \cos 2\pi f_m t]$ which is, by definition, 2π times the instantaneous frequency. Frequency modulated signals can be produced in several ways. In the microwave band it is usual to generate the signal with a *reflex klystron* with the modulating signal on the repeller. Demodulation is accomplished with a discriminator after amplifying and hard limiting the signal to remove any accidental amplitude modulation. If $\Delta F/f_m$ (called the *modulation index*) is large, we describe the signal as *wide band* FM. In that case the r.f. signal occupies more bandwidth than the original intelligence. However, there is an important advantage. So long as the signal to noise ratio is 6 dB or greater, a very significant noise suppression occurs on demodulating wide band FM signals. *This noise reduction is proportional to the square of the modulation index.* However, that is not all gain, because if the noise per unit bandwidth is constant (white) the noise in the r.f. channel required to carry the intelligence is proportional to ΔF. Therefore as we increase the modulation index used to transmit a particular bandwidth of intelligence, we gain linearly in output signal to noise ratio although the r.f. signal to noise ratio decreases - up to the point where the r.f. signal to noise ratio becomes too low to consistantly capture the discriminator (about 6 dB). Below that point we will begin to get signal dropouts, i.e., periods of time when all intelligence is lost.

Pulse code modulation, abbreviated PCM, is a technique which, like broadband FM, accomplishes a great *increase in signal to nosie ratio at the expense of utilizing additional bandwidth.* Essentially it measures the value of the signal voltage at points spaced 1/2B apart (where B is the intelligence bandwidth) and converts the value to a binary number which it transmits in digital form sequentially. Thus if each voltage value is encoded as a 4 digit binary number, and transmitted as a sequence of on and off, we will require 4 times the bandwidth, but baring errors in recognizing the presence or absence of the signal, the receiver will always have the correct voltage value to 1 part in 32 or 3%.# Clearly more digits could be used, *parity checks* could be added to detect errors, etc. In PCM there are two distinct sources of noise. One is due to *system errors* in deciding whether or not there is a pulse present at a particular time; the other is due to *quantization.* The signal to quantization noise ratio is 2^{2n} which for the 4 binary digit code chosen above would be 256/1 or about 24 dB. (Ref.40) The decision error noise would, of course, reduce this signal to noise ratio, but that depends on the r.f. signal to noise ratio in a fairly complex way.

So far we have assumed that the r.f. bandwidth is at least as great as the intelligence bandwidth, and in fact all common communication systems work in that way. However, that is not necessarily so. If the *signal to noise ratio is adequately high, some of it can be traded for bandwidth.* For example, if we consider the signal as a set of values 1/2B seconds apart, we could encode the values in pairs, for example, according to the following scheme. First we quantize each value in n steps so that a pair of adjacent values can be represented by a pair of numbers i, j where $0 < i < n$, $0 < j < n$. Then we generate a voltage of value $(n+1)i+j$. Thus, for example, for the case of n=9; i=8, j=4 would be represented by a value of 84, i=1 j=9 would be represented by 19, etc. To reconstruct the original signal without error, we would have to be able to measure the encoded pulse heights to an accuracy of 1% instead of the 10% required if we did not use this scheme for reducing bandwidth. This will require a very high signal to noise ratio. However, it demonstrates that such a tradeoff is in principle possible.

Shannon (Ref.6) has shown that when the noise in the system is Johnson noise, or noise having equivalent statistical properties, the capacity of a channel for transmitting information, measured in bits per second, is

$$\text{Maximum Ideal Capacity} = B \log_2 (1 + \eta) \qquad (7.4)$$

where B = bandwidth in Hz

$\log_2$ = log to the base 2

η = signal to noise power ratio.

Thus if we were sufficiently ingenious in designing our method of encoding and decoding the signal, we should, in principle, be able to transmit that

---

#i.e., 3% of the maximum value if a linear amplitude division is used. One can also digitize and transmit the log of the amplitude, or any other function.

much error free intelligence. However, it is not always clear how to do it, and even if it were, the amount of equipment and its complexity might be excessive. We rarely, in practical cases, even approach that limit. For example, at a signal to noise ratio of 15, or about 12 dB, Eq.7.4 predicts that the capacity in bits per second should be 4 times the bandwidth. In usual practice it is equal to the bandwidth or less.

7.2 Multiplexing.

There are two practically important methods of combining and sorting out separate intelligence signals on a single channel. These are called *time division multiplexing* (TDM) and *frequency division multiplexing* (FDM). In TDM, time slots are allocated to each intelligence source and each is sampled in turn and the sampled value is transmitted so that, for example, if 3 intelligence signals are to be combined, the combined signal contains in sequence sampled values from 1,2,3,1,2,3...., and the receiver is keyed to the sequence and sorts them out. In FDM each intelligence signal is shifted in frequency by a different amount so that there is no frequency overlap and the signals are linearly combined and transmitted. The receiver sorts them out by filtering.

The quantitative features of TDM are related to the fact that a sharp narrow pulse filtered to exclude all frequencies above some maximum value F_1 will resemble Fig.7.2 in the time domain.

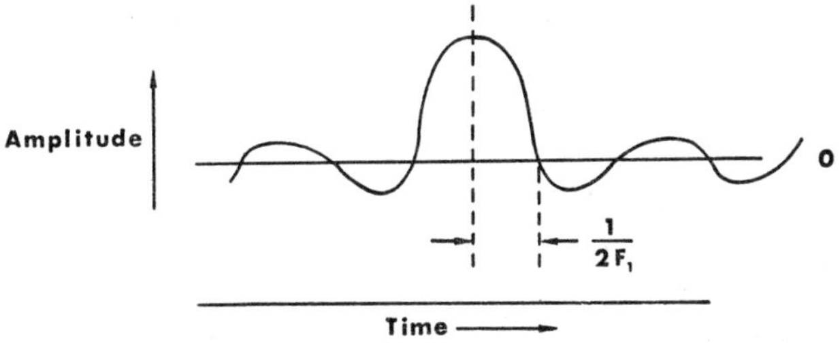

Fig.7.2. Shape of sharp pulse after filtering out all frequencies greater than F_1.

Thus, if we take an intelligence signal, sample it to obtain a flat topped steep sided pulse, and filter the resulting sample pulses to exclude all frequencies greater than F_1, each pulse will resemble Fig. 7.2 in shape, with the first axis crossing separated from the peak by $1/2F_1$ seconds and other axis crossings separated from each other by $1/2F_1$ seconds. If we have n samples per second to transmit, we can place each sample peak at the axis crossing of all other samples by sampling at intervals of $1/2F_1$ seconds. Thus if there are n samples, each is sampled at intervals of $n/2F_1$ seconds. After transmitting the signal we sample the received signal, sort out the samples into n channels and filter each to bandwidth F_1/n. If everything is done perfectly, there will be no crosstalk between channels, and the original intelligence will be completely recovered.

The quantitative features of FDM are somewhat more obvious. In addition the circuitry is simpler. TDM is rarely used except where it lends itself readily to switching circuits used to gather intelligence sequentially from several sources. FDM is widely used in telephone circuits.

7.3 Handling of Voice Communication.

The normal voice channel bandwidth used in telephone communications is 3.2 kHz. In addition, it is normal practice to use an 800 Hz guardband between channels, *resulting in a* 4 kHz *spacing between channels.*

The usual means of transmitting voice communications by microwaves is by FDM-FM or frequency division multiplex-frequency modulation. The frequency division multiplexing is accomplished by using single sideband amplitude modulation for each voice channel and separating the carriers by 4 kHz. *The combination of many such voice channels makes up the baseband.* In the typical multiplex grouping of 12 channels, the carrier frequency for the bottom channel is 12 kHz (the lower frequencies are reserved for the order wire which carries switching signals). If there are 12 channels so frequency multiplexed therefore, the top frequency carrier will be at 56 kHz. This complex baseband signal which covers 48 kHz, when carrying voice communication channels, appears noiselike. For transmission over a microwave link it is used to frequency modulate the microwave carrier signal. The modulation index is ideally of the order of 100, but it is variable, depending on how many of the available channels are occupied, the level of speech, etc. More than 12 voice channels may be used to make up the baseband. Typically several 12 voice channel basebands are combined by shifting each 12 channel group by 60 kilohertz before combining them to form a single wide frequency baseband. *A 60 channel supergroup, carrying 5 such channel groups* is normal for long range transmissions and even larger groups are used, e.g., 60 such groups are combined to form a supergroup for INTELSAT IV transmission. After the signal is transmitted and received, a discriminator is used to recover the baseband signal which then passes through a series of 4 kHz bandwidth filters to separate individual voice channels.

The signal level in a telephone channel is standardized at *1 milliwatt at 1000 Hz at the receiving instrument.* The normal design noise level is 50 dB below that level or 10^{-8} watts, which can be exceeded by about 8 dB for a small fraction of the time when circuits are heavily loaded. Thus, at the receiving instrument the signal to noise (assuming the signal to be at its standard level) is 50 dB. This seems quite high, but it is necessary to keep the noise from being plainly audible, because of the great dynamic range of the human ear, particularly during breaks in the speech.

This tolerable noise level at the receiver is budgeted among many sources including intermodulation among channels due to nonlinearity in the system and system overload. However, when microwave links are involved, the major part of the noise budget *(about 80%) is allocated to thermal noise.* This does not mean, however, that the signal to noise level on the microwave link must be 50 dB. It need not be anywhere near that high because as noted in Section 7.1 the wideband FM system has the characteristic of suppressing the noise by the square of the modulation index if the carrier to noise ratio (designated as C/N) is at least 6 dB. The normal design value is about 10 dB for carrier to thermal noise ratio at the microwave receiver output. The 40 dB improvement is characteristic of wide band FM systems where the peak frequency deviation of the FM signal is 100 times the highest modulation frequency. Thus the carrier to noise ratio (C/N) is designed to provide about four dB margin for system degradation. For the INTELSAT IV system, for example, the up and down links combined are designed to produce less than 7.9×10^{-9} watts of noise at the telephone receiver.

7.4 Communications Satellites.

The first earth satellite, SPUTNIK I, was launched by the Soviet Union in October of 1957. The first U.S. Satellite, SCORE, launched in December 1958, was possibly the first communications satellite since it relayed messages as it orbited the earth. The first repeater satellite was TELSTAR I launched in July 1962. In July 1963 the first successful synchronous orbit communication satellite was placed over the Indian Ocean. It was used for military purposes.

At present the commercial communications satellite field is dominated by the Communications Satellite Corporation (COMSAT) with its INTELSAT series of satellites. The first of these was EARLY BIRD launched in April, 1965. As of 1976 about 80 countries have earth stations communicating with the INTELSAT satellites which are in *synchronous orbit* 22,300 miles above the equator.

The height of any circular satellite orbit is determined by a balance between centrifugal and gravitational force, i.e.,

$$\omega^2 r = 32.2 \left[\frac{R}{r}\right]^2$$

where r = radius of the orbit

R = radius of the earth (4000 miles)

32.2 = acceleration of gravity at the earth's surface (which falls off as the square of the distance from the earth's center)

ω = 2π times the number of revolutions per second.

For a synchronous orbit, (1 revolution per day)

$$\omega = 2\pi \times \frac{1}{24 \times 60 \times 60} \times \frac{365}{364} = 7.30 \times 10^{-5} \text{ radians per second}$$

(where we have corrected for the sidereal day)

Solving Eq.7.5 with this value of ω gives r = 26,300 miles, and since the earth's radius is 4,000 miles, the height above the earth's surface is 22,300 miles.

The most recent COMSAT series of satellites for commercial communications is the INTELSAT IV series. The cost of this program is estimated at $250M for satellites and putting them into orbit. This does not include the earth stations. 10 satellites are involved, several of which are standby-backup.

The uplink frequencies are between 5.925 and 6.425 GHz and the downlink frequencies are between 3.7 and 4.2 GHz. The total electrical power of the satellite is derived from solar cells with batteries used to store power for the periods during which the satellite is eclipsed from the sun. The uplink power is of the order of kilowatts, provided by normal earth electrical power sources.

The INTELSAT IV satellite weighs 3000 lbs and is 17 feet high and 8 feet in diameter. It has twelve 36MHz channels, each capable, under optimum conditions, of transmitting color TV or 750 telephone circuits (3.6MHz baseband). The total usable r.f. communications bandwidth is 432 MHz. Note that the r.f. bandwidth under these optimum conditions is only 10 times the voice bandwidth rather than 100 times as mentioned in section 7.3 above. If the modulation index ($\Delta F/f_m$)is 10, the improvement in S/N on demodulation will be about 20 dB. However, the situation is not really that bad because all the voice channels are not filled to capacity at the same time, e.g., there are pauses in conversation, lower than maximum level voices, etc. The system accomodates to this reduced modulation by increasing the modulation index to fill the r.f. band. This total capacity of 9000 telephone circuits (12 x 750) is a maximum, and can be used only when the system has a very good r.f. signal to noise ratio.

Each satellite has 12 communications repeaters and a combination of *wide area (earth beam) and steerable spot beam antennas*. The spot beam antennas are used only for downlink transmission. They are 50 inches in diameter and are pointed at high communication density areas such as the East Coast of the United States. By virtue of this large size, they achieve high gain, but cannot cover a wide area (see Eq.4.15). The earth beam antennas cover the entire sector of the earth which is visible from the satellite, but have lower gain. Because of the higher gain of the spot beam antennas, the performance of the system is best when most or all of the traffic is from the high population density areas served by these antennas. It is under these conditions that the 9000 voice channel maximum and the modulation index less than 100, mentioned in the previous paragraph can be used.

The satellites are spin stabilized but the entire communication system, transponders and antennas, are counter rotated so that they are not spinning. The position and orientation of the satellite can be adjusted at frequent intervals by means of small jets and command signals which are provided by pulse code modulation. The system is capable of receiving over 200 types of commands, 60 of which are concerned with controlling the position and aspect of the satellite while the remaining commands are concerned with selecting multiple operating modes of the communications system.

As noted, electrical power for the satellite comes from *solar cells*.

The sun delivers about 1 kilowatt per square meter of surface in the vicinity of the earth and the solar cell efficiencies are about 10%. Most of the satellite surface is covered by solar cells. The main r.f. power source in the INTELSAT IV satellite comes from its *12 traveling wave tubes, each of which has a 6 watt nominal output.* The product of the radiated power times the antenna gain called the *equivalent isotropic radiated power (EIRP) is 22 dBw* for the global antennas which have a beam width of 17°; and 33.7 dBw for the spot beams which have a beam width of 4.5°. The allowance for losses in the transmission system is about 4 dB. Spot beam antennas are not used to receive uplink signals because the high powered transmissions from the earth make them unnecessary. The gain of the spot beam antennas is about 30 dB, and that of the global antennas about 18 dB. The system includes two global receive antennas, two global transmit antennas, and two spot beam transmit antennas. In addition, there are smaller telemetry antennas. The receive polarization at the satellite is left hand circular and the transmit polarization is right hand circular. The system incorporates a great deal of redundancy, using wide band ferrite switches to route the signals to and from the receivers and traveling wave tubes which are in use.

In a frequency modulation system the instantaneous frequency is proportional to the voltage which is impressed upon the modulator. When a large number of voice channels are combined to generate the modulation signal, the character of the signal is noiselike, and can be described only statistically. The statistics depend upon the detailed character of the speech, the loudness, the length of pauses, and many other characteristics, but in general the combined signal will have peak values which are many times as great as its average value. Further, as more and more voice channels are added to achieve a total of 750 for each of the INTELSAT IV r.f. channels, these peak values will increase. The system cannot be designed to accomodate the extremely infrequent maximum peaks which may occur since each of these peaks represents an extreme frequency deviation of the system which might drive it out of its channel. If 1% of these peaks are clipped, it has been found that no serious degradation in speech quality results. However, if very much more than 1% of them are clipped, there is a very noticable degradation in speech quality. There are *peak limiting circuits which hold this clipping to 1%* or less by varying the modulation index when the modulating power becomes large due to a large number of voice channels and loud voices. This variation in modulation index is a slow action somewhat like an automatic volume control. *Changes in modulation index need not be tracked* at the other end of the circuit because its only consequence, when it passes through the demodulator, is a change in loudness which is uniform on all the channels and compensated for by automatic volume control or other means on the receiving channels. It does not affect the recovery of the individual voice channels in the baseband because this is done by filters operating on the recovered baseband frequency.

Peak limiting does, however, have a penalty in signal to noise ratio by virtue of the fact that the reduction in modulation index which it imposes when all voice channels are loaded, reduces the signal to noise improvement which occurs on FM demodulation.

The earth stations employ antennas as large as 100 feet in diameter, and transmit powers of the order of kilowatts. The important characteristic of the earth station is its signal receiving capability expressed in terms of *G/T which is actually the antenna gain in dB minus the effective temperature of the receiver and antenna expressed in terms of dB relative to 1°K.* Thus, for example, a 60 dB gain antenna with an effective receiver temperature of 100°K would have a G/T of 40 dB. The temperature includes the sky temperature which is typically from 26°K at a 5° elevation, to 4°K at 90° elevation. Typical G/T values for earth stations are of the order of 40 dB. The International Telecommunications Satellite Consortium, the governing body of INTELSAT, requires that at 5° elevation under clear sky conditions, G/T must be greater than 39 dB at 4GHz and the design value must be greater than 40.7 dB for the earth station to be approved as part of the INTELSAT IV system. Also, the antenna gain must be greater than 57 dB at this frequency.

The earth transmitting stations must deliver 61-63.5 dB$_\omega$ E I R P per voice channel if there are less than 12 voice channels. For larger numbers of channels, the power level per channel is reduced by an 'activity factor' which varies from 85% for 12 channels to 57% for 60 channels. The total transmitted power must by 74 to 98 dB$_\omega$ E I R P, depending on bandwidth and other factors. The level must be held to within .5 dB of a value called for at any particular time by the satellite system manager. (Ref.33)

Satellite communication links are substantially independent of atmospheric conditions because of the low frequency and because the atmosphere is fairly thin and uniform above a height of 30,000 feet. The ionosphere has no significant effect at the high frequencies which are used. In a heavy rainstorm, however, there may be 1 to 2 dB of attenuation introduced at the downlink frequencies when the antenna elevation angle is low. In accordance with the discussion in Section 6.4, the absorption will be about 6 times greater at the higher uplink frequencies. However, attenuation in the uplink is much less important because the power available for transmission is so much greater. In fact, it is the greater sensitivity of the downlink to losses coupled with the fact that the water droplet losses increase with frequency, which dictates that the satellite downlink frequencies be lower than the uplink frequencies. (See, for example, Ref.34.)

The problem of *delay* in satellite communications is marginally serious. Since the distance from earth to satellite and back is 45,000 miles, the time required for a signal to make the round trip is almost 1/4 second. When one person stops talking, if the other party answers immediately after he hears the first person stop, the first person will experience a delay of almost 1/2 second before he hears the reply, because of the round trip travel time. When satellite voice communications were first employed, the engineers were surprised that the callers did not object to this delay.

A second problem associated with this delay is *echo sensitivity*. One can tolerate fairly large echoes if the delay is short. For example, the echo and reverberation in a normal room is quite loud but not troublesome. However, if the delay is long, e.g., in a large room, even a small echo can be quite disturbing. There is normally an echo in a telephone line due to missmatch at the receiver and other factors. When satellite relays were employed, these echoes became much more troublesome because of the longer delays involved. Special echo suppressor circuits must be used.

In the future, as satellite costs are reduced, we can look forward to the use of satellites in domestic telephone service as well as in intercontinental service. Preliminary plans are already underway for such service (Ref.35). However, *if two satellite hops are used in series* in a telephone link, the delays will surely be unacceptable. There are also some serious problems in frequency allocation, and it is planned to use the same band now used for terrestrial link (TL) relays. This is about *10 GHz* and since absorption by water is serious at these frequencies, *site diversity* will probably be required, i.e., there will be alternate receiving locations, perhaps 20 miles apart, cabled together so that if it is raining very hard at one site, the other can be used.

There are three unique *types of outages* in satellite communications systems. The first of these is solar eclipse, when the satellite passes through the shadow of the earth, causing its solar cells to be inoperative. This occurs for one hour on each of 43 consecutive nights in the spring and fall. Battery backup is used during these periods. The second unique outage is solar glare, when the sun is directly behind the satellite so that a receiving antenna pointed at it will receive thermal noise from the sun, swamping the satellite signal. This causes an outage lasting about 10 minutes, and generally occurs on 5 successive days twice a year at times depending on the earth station location. The third unique outage occurs when aircraft fly across the communication path. This is a rare phenomenon for most stations (Ref.36).

7.5 The TL Radio Relay System.

The TL radio relay system (Ref.37) is a system of short hops (usually less than 30 miles) which operates in the frequency range of 10.7 to 11.7 GHz. Most of the recently installed or modified TL relay stations employ periscope antennas, and you have probably seen these on the top of tall buildings. The important characteristic of the TL system which makes it economically viable as compared to the alternative of cables and longer microwave links is that the relay stations are unmanned and have adequate reliability to operate with only an occasional service call or maintenance check. A relay station requires only 170 watts of power and employs a continuously charged battery power supply which has sufficient reserve to carry the system for 20 hours under average ambient conditions in the event the a.c. power fails. The r.f. power output tubes are klystrons with an output power of 100 milliwatts.

Almost every other component in the relay station (except for the local oscillator tubes) is solid state.

The *order wire and alarm system* of the TL radio relay stations are moderately complicated because they must convey considerable information about the state of the relay stations, since they are unmanned. For example, they must sound an alarm when the a.c. power fails even though the battery is carrying the load. They must convey information about failure of the lightning arrestor system and even the aircraft warning lights. The order wire signals are carried over a line of relay stations and returned over the same links from the station which is being interrogated so that the continuity of the order wire can be checked.

The band from 10.7 to 11.7 GHz is divided into 24 r.f. channels, each about 40 MHz wide. In a given relay link only 12 of these are used, resulting in 80 MHz spacing between mid-channel frequencies. 6 of these r.f. channels are used for transmission in each direction. Further, as shown in Fig. 7.3, when a channel is used at a given relay station, adequate frequency separation between transmitters and receivers is achieved by using the upper half of the band for transmitting and the lower half for receiving or vice versa. Thus each repeater section uses half of the available 1 GHz bandwidth and the next repeater section uses the other half. The problem of *overreach*, e.g., the problem of relay station 4 receiving signals sent from relay station 1 to relay station 2 (since they utilize the same frequencies) is solved by reversing the direction of polarization, vertical versus horizontal, the next time the same frequency allocations are used in a repeater section, and using angles other than 180° between paths.

Each channel (40 MHz wide) carries 6 groups of voice channels using FDM-FM similar to that described in Section 7.3 above (72 voice channels). The modulation index is about 100 and peak limiting is used. The principal problems in the TL radio system are atomspherics. Protection against rain attenuation must be assured by engineering sufficient fading margin into the system and by using path lengths appropriate to the particular area of the country. Path lengths range from 10 miles or less in the heavy rain areas to 30 miles in the dry areas. The antennas are generally 5 or 10 feet in diameter. The 5-foot antenna has an effective area of about 1 square meter after allowing for an antenna aperture efficiency of about 60%. This gives a gain slightly over 30 dB. Thus over a 20 kilometer span one can expect a received signal level of about 2×10^{-8} watts. The receivers are not particularly low noise configurations and may have noise figures as large as 13 dB, resulting in equivalent input noise levels about 3×10^{-12} watts in a 40 MHz band. However, this still gives an r.f. signal to noise level of the order of 40 dB which is necessary because rain attenuation and fading may provide as much as 30 dB loss under unfavorable conditions, and the noise will accumulate as the signal is transmitted over successive hops.

Fading in the TL radio relay system and in longer hop terestrial links, is a phenomenon which is not completely understood, but its general origin is in the bending of the radio paths described in Section 6.2. If the normal downward refraction of the path is greatly decreased or even reversed due to temporary atmospheric conditions, usually humidity gradients, there may no longer be a direct path between the antennas which is free of physical interference. Sometimes the direct path will persist, but there will also be a ground reflected path at a small grazing angle which arrives out of phase with the direct path. Because the reflectivity of the ground is quite high at small grazing angles, this reflected signal arrival may be quite close in magnitude to the direct arrival, resulting in a very small net signal. Apparently there are also other causes of fading including path bending in the horizontal plane and even path splitting.

Many of the installations include *frequency diversity* systems to provide against service interruptions caused by multipath fading and equipment failures. One advantage of this is that alternate facilities are then automatically available during maintenance periods. With diversity these periods can be scheduled at convenient times. In the diversity type system the same baseband signal is used to modulate 2 different 40 MHz channels widely separated in frequency and using opposite polarizations. The control signal is a 2600 cycle per second pilot continuously transmitted at both frequencies. Upon receipt of the microwave signals at the next relay station, a comparator circuit compares the received level of this 2600 cycle per second signal on the two bands and controls a trigger circuit which activates a relay to se-

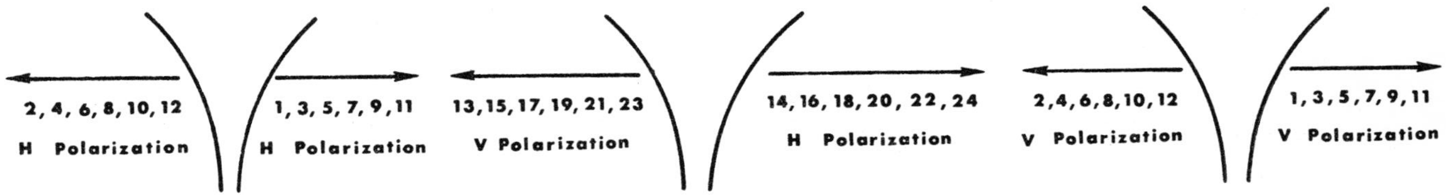

FIGURE 7.3 -- Scheme of TL radio link.
Numbers indicate rf channels in use. Each cnannel is 40 mHz wide.

lect the channel which has the stronger signal.

The i.f. frequency is standardized at 70 MHz and at each installation the received signal is brought down to the baseband signal, at which point channels originating in the local area are removed or added before using the baseband to modulate the r.f. for the next hop.

The TL radio relay system is primarily intended for short haul service, 200 to 300 miles at the maximum. Occasionally, however, telephone transmissions all the way across the country may be routed through these short haul lines. When this occurs there may be some significant degradation in the quality of transmission because of the fact that the signal has gone through so many relay links (often several hundred). The degradation, of course, is cumulative since any noise or cross modulation which enters the communication channel at any point cannot be removed by subsequent links.

7.6 Radar.

The basic *radar equation*, arranged so that it can be understood functionally is

$$P_r = \frac{P_t G}{4\pi R^2} \sigma \frac{A_e}{4\pi R^2} \qquad (7.6)$$

where P_r = received power

P_t = transmitted power

G = gain of the transmitting antenna

R = range

σ = cross section of the radar target (see Section 6.3)

A_e = effective area of the receiving antenna.

With rare exceptions, the transmitting antenna and the receiving antenna are the same. The signal consists of short pulses and the antenna is used for reception of echoes between pulses. Therefore Eq.4.3 can be used to replace either G or A_e so that both do not appear. The equation may, of course, be solved for R, and if P_r is replaced by the minimum detectable signal (S_{min}) the value of R so found will be the *maximum range of the radar* (as a function of σ).

If we assume coherent detection processing, we can write the *minimum detectable signal* in the form

$$S_{min} = d\ F\ kT_0B/Bt \qquad (7.7)$$

where d = detection index required (see Eq.3.7 and Fig.3.1 and 3.2)

F = noise figure of the radar receiver (see Eq.1.21)

k = Boltzman's constant (1.38×10^{-23} watts seconds/degree)

T_0 = 290°K (The input noise is usually kT_0B. If not, FkT_0B must be replaced by the actual noise out of the receiver divided by its total cascaded gain.)

B = bandwidth

t = duration of the pulse (Bt is the coherent processing gain).

Solving Eq.7.6 for R_{max} with $P_r = S_{min}$ given by Eq.7.7 and $A_e = \lambda^2 G/4\pi$ results in

$$R_{max} = \left[\frac{E_t\ G^2 \sigma\ \lambda^2}{(4\pi)^3\ d\ F\ k\ T_0\ L}\right]^{1/4} \tag{7.8}$$

where $E_t = P_t\ t$ = total energy transmitted per pulse

λ = wavelength

L = term inserted to account for losses and departure from ideal performance.

The value of d can be selected on the basis of false alarm and missed detection requirements using Fig.3.2. The noise figure F, the radar cross section σ, and losses can then be estimated, and one then has a design relationship between E_t, G, λ and R_{max}.

In the *search mode* it is important that a radar cover a particular sector or solid angle in some length of time, called the *sector scan time*. The solid angle covered by each pulse is approximately related to the gain of the antenna by $\Omega = 4\pi/G$ according to Eq.4.14. Therefore, the total solid angle which can be covered in t_s seconds is given by

$$\Omega_t = 4\pi n\ t_s/G \tag{7.9}$$

where n = number of pulses per second.

If we express G^2 in Eq.7.8 in terms of GA_e and substitute for G from Eq. 7.9, we obtain the radar search equation

$$t_s = \frac{F\ k\ T_0\ L}{P_{av}} \cdot \frac{d\ R^4_{max}}{\sigma A_e}\ \Omega_t \tag{7.10}$$

where t_s = the search time

P_{av} = average radar power ($n\ E_t$)

Ω_t = solid angle to be searched.

One significant thing about Eq.7.10 is that the search rate of this coherent processing radar depends on the average transmitted power, the effective area of the antenna and only a small number of other factors. The second noteworthy thing is that the search time is inversely proportional to the effective area of the antenna. Large effective areas are easy to achieve at low frequencies, even with fairly small antennas, and it is for this reason that search radars generally operate at low frequencies. On the other hand, radars used for fine localization require narrow beams, which are most easily achieved at high frequencies. This treatment assumes coherent processing, but if the bandwidth-time product is small, the difference between coherent and incoherent processing is minor. This is the usual case in radars.

Clutter in a radar is unwanted echoes from the ground and inhomogenities in the atmosphere. The echo must be detected in the presence of this clutter as well as noise. There are two important techniques for doing this. The most common of these is to *use short pulses to minimize clutter*. The only environmental echoes which arrive at the same time as the target echo are those coming from the same range plus or minus half the pulse length (the physical length of the pulse as it travels through space). By shortening the pulse, the competing clutter is reduced. Alternatively, for moving targets such as missiles or aircraft, *doppler discrimination* can be used. Since most of the objects generating clutter are relatively stationary, the target echo can be recognized by its doppler shift. However, the pulse must be long if the receiver is to be able to recognize small doppler shifts. The doppler shift is

$$\Delta F = \frac{2v}{c} f_0 \qquad (7.11)$$

where v = target velocity

c = 3×10^8 meters/second

f_0 = radar frequency.

Thus for a 1000 MHz radar and a 200 knot aircraft, the doppler frequency will be about 600 Hz and if the pulse is shorter than about 2 milliseconds, it will be difficult to measure its frequency. A 2 millisecond pulse is quite long for radar and will result is quite poor range resolution. Pulses of a microsecond or less are more usual, and necessary for short range radars so that the transmitted pulse will end before the echo starts to arrive. In fact a 2 millisecond transmitted pulse will blank out echoes from targets within 200 miles. Of course, higher aircraft speeds and higher radar frequencies will increase doppler shift and permit shorter pulses.

Aside from such minimum range problems, the requirements of doppler discrimination and the desire to minimize the clutter and obtain range resolution by using short pulse are mutually opposed. However, a compromise can be reached by using *pulse doppler techniques*. Short pulses are used but phase coherence is maintained between pulses so that the doppler can be resolved by comparing the phase of successive echoes. There are, however, complicated possibilities of ambiguity in pulse doppler.

Another problem with short pulses is illustrated by Eq.7.8 where it is shown that the maximum range of the radar depends on the energy per pulse. To put a lot of energy in a short pulse requires high peak power, which may, in some cases, encounter component and circuit limitations. Such problems can often be surmounted by *pulse compression techniques*. Suppose, for example, that the pulse varies linearly in frequency from high to low during its "on" time. On receipt of an echo a circuit can be used to delay the signal in proportion to its frequency so that the initial high frequency part of the echo arrives at the circuit output at the same time as the later, low frequency part; and all parts arrive simultaneously. In most respects, such as detection range, this pulse compression technique is equivalent to using a shorter, higher power pulse. The technique is an example of a matched filter (see Section 3.3).

Most simple radars employ a *PPI or Plan Position Indicator display*. This is a circular oscilloscope face on which range and bearing are displayed for each echo and clutter element. The oscilloscope beam is intensity modulated by the echoes as it sweeps outward from the center of the display after each transmitted pulse,on a bearing corresponding to that of the antenna. In that way as the antenna rotates it paints a picture of all objects in the vicinity of the radar. The scope phosphor is selected to provide suitable optical emission decay rates so that the entire picture, or substantial portions of it, can be viewed simultaneously by the operator. However, the third dimension, i.e., height, cannot be displayed in so simple a way. This is a particular problem for airport radars although some partial solutions are available.

An important recent development in military radars is the *phased array antennas* which generally consist of a planar array of dipoles which do not physically move, but direct their beam by proper phasing of the power fed to them. The virtue of this arrangement is that the transmit and receive beams can be moved much more rapidly than they could move mechanically. This is particularly valuable in *track while scan radars* which can have several pulses on their way in widely different directions in accordance with computer instructions, and then direct the receiving beam to each direction to listen for returning echoes at precisely the time (between subsequent transmit pulses) when such echoes are expected.

Weather radars generally operate in the frequency range from 3-10 GHz so that they receive detectable echoes from rain and clouds but the signals are not disasterously attenuated by passage through them. The longer range weather radars operate in the lower part of this frequency range. *Ship navigation radars* generally operate at frequencies down to 100 MHz. They have no need for very high range or angular resolution or for great sensitivity since they are concerned with targets of large scattering cross section. They generally perform a continuous horizon search, presenting the output on a PPI.

It is *missile defense radars* which are currently pushing the state of the art. Long ranges are required for adequate warning times, accurate localization is required for defense, and the target cross sections are small. It is for this application that phased array antennas were developed. Sophisticated computerized echo analysis techniques are employed, and high powered tubes and other components are developed for this application.

CW radars (Ref.39) which transmit continuous single frequency signals are used for a number of purposes related to doppler measurement. Such radars generally use separate transmitting and receiving antennas in close proximity to each other. There is considerable leakage between them, and the direct leakage signal into the receiving antenna acts as local oscillator signal with the i.f. frequency being the doppler due to target motion. Such receivers are called *homodyne receivers*. The quantity of interest is usually the doppler frequency which can be read on a meter or detected aurally. Such radars are used by the police in automotive speed control and in doppler navigation by aircraft.

Frequency modulated CW radars are very much like the CW radars discussed in the paragraph above except that the transmitted frequency is slowly swept in one direction. Therefore the echo, even from a stationary object, will be at a different frequency from the signal being transmitted at the instant when the echo is received. Therefore, there is a small i.f. frequency which is larger for more distant objects. This is the principle of most *radar altimeters*.

CW or FM CW radars are simpler than pulsed radars, and their receivers are particularly simple because they need no local oscillator signal other than transmitter leakage. However, because of the very small i.f. frequency they suffer from *flicker noise*. This is a type of noise prelevant in semiconductor devices and vacuum tubes which varies with frequency approximately as 1/f. In most cases the power levels, ranges, and performance requirements are such that flicker noise can be tolerated, but it can be avoided by adding a separate local oscillator and an i.f. amplifier at a frequency of a few megahertz before developing a doppler frequency with a second detector.

REFERENCES

1. Johnson, J.B., Thermal Agitation of Electricity in Conductors, Physical Review, 32, 97-109 (July 1928).

2. Nyquist, H., Thermal Agitation of Electric Charges in Conductors, Physical Review, 32, 110-113 (July 1928).

3. Mumford, W.W., and Schelbe, E.G., Noise Performance Factors in Communication Systems, Horizon House, 1968.

4. Burrington and May, Probability and Statistics, McGraw-Hill, 1958.

5. Feller, William, An Introduction to Probability Theory and Its Applications, John Wiley and Sons, 1957.

6. Shannon, C., Communication in the Presence of Noise, Proc. IRE, 37, 10-21, (January 1949).

7. Feller, William, Loc.Cit. IX.5.

8. Ibid. X.1.

9. Viterbi, A.J., Principles of Coherent Communication, McGraw-Hill, 1966, Section 1.3.

10. Goldman, Stanford, Frequency Analysis Modulation and Noise, McGraw-Hill, 1948.

11. Blake, Lamont V., A Post Detection Method of Measuring Prediction RF Signal-to-Noise Ratio, IEEE Transactions on Aerospace and Electronic Systems, AES-8, No. 2, (March 1972).

12. Selected Methods and Models in Military Operations Research, Naval Postgraduate School, Monterey, California, 1971. Superintendent of Documents, U.S. Government Printing Office, Stock No. 0851-0054.

13. Davenport, W.B., and Root, W.L., Introduction to Random Signals and Noise, McGraw-Hill, 1958.

14. Skolnick, Merrill, Introduction to Radar Systems, McGraw-Hill, 1971, Section 8.6.

15. Peterson, W.W., and T.G. Birdsall, The Theory of Signal Detectability, Univ. Mich. Eng. Res. Inst. Rept. 13, 1953.

16. Skolnick, Merrill, Loco.Cit., Section 7.2.

17. Ibid., Section 7.5.

18. Ibid., Section 7.3.

19. Ramo, Simon, and Whinnery, John, Fields and Waves in Modern Radio, John Whiley and Sons, 1954.

20. Skolnick, Merrill, Loco,Cit., Section 7.2.

21. Silver, Samuel, Microwave Antenna Theory and Design, McGraw-Hill 1972.

22. Breit, G., and M.A. Tuve, A Test of the Existance of the Conducting Layer, Phys.Rev., 28, 574-575 (September 1926).

23. Hewish, A., S.J. Bell, J.D.H. Pilkington, P.F. Scott, R.A. Collins, Nature 217, 709 1968.

24. Drake, F.D., E.J. Gundermann, D.L. Jauncey, J.M. Comella, G.A. Zeissig, and H.D. Craft, Jr., The Rapidly Pulsating Source in VULPECULA, Science, 160, No. 3827, 503-507 (May 1968).

25. Davies, Kenneth, Ionospheric Radio Propagation, U.S. Department of Commerce, National Bureau of Standards Monograph 80, 1965.

26. Adler, R.B., Chu, L.J., and Fano, R.M., Electromagnetic Energy Transmission and Radiation, John Wiley and Sons, 1965, Section 7.4.

27. Urick, R.J., Principles of Underwater Sound for Engineers, McGraw-Hill, 1967.

28. Barton, David K., Radar Systems Analysis, Prentice-Hall, 1965, Section 15.3.

29. Skolnick, Merrill I., Loco.Cit., Section 11.4.

30. Lenkurt Demodulator,General Telephone and Electronics, 1105 County Road San Carlos, California, 94070, July and August 1972.

31. Skolnick, Merrill I., Loco,Cit., Section 2.7.

32. Lenkurt Demodulator, Loco.Cit., April 1969.

33. Interim Committee on Satellite Communication ICSC-45-13E W/1/70, Performance Characteristics of Earth stations in the INTELSAT IV System, 1970.

34. Skolnick, Merrill I., Loco.Cit., Figures 11.11 and 12.11.

35. The Lenkurt Demodulator, Loco.Cit., July 1970.

36. The Lenkurt Demodulator, Loco.Cit., June 1970.

37. Hathaway, S.D., Sagaser, D.D., and Word, J.A., Bell System Technical Journal, XLII, 5, 2297-2353 (September 1963).

38. Skolnick, Merrill I., Loco.Cit., Section 2.1.

39. Ibid., Chapter 3.

40. Pierce, John R., Electrons, Waves and Messages, Hanover House, Garden City, New Jersey, 237 (1956).

APPENDIX A

IMPORTANT FORMULAE, NUMBERS AND CONCEPTS

I - THERMAL NOISE

$$V_{r.m.s.} = \sqrt{4RkTB}$$

k = Boltzmann's constant (1.38×10^{-23} watt-second/degree)

The available thermal noise power from any resistor is kTB.

The transmission bandwidth of a lossless filter or 2 port must be the same in both directions.

The noise figure $F = \frac{(S/N)_{in}}{(S/N)_{out}}$ (if $N_{in} = kT_0 B_n$)

$$F = 1 + \frac{T_e}{T_0}$$

$$\text{noise out} = k B_n G (T + T_e).$$

For an attenuator $F = 1 + (\alpha - 1) T_1/T_0$

$$T_e = (\alpha - 1)T_1$$

For a cascade of components $T_e = T_{e1} + \frac{T_{e2}}{G_1} + \frac{T_{e3}}{G_1G_2} + \ldots$

$$F = F_1 + \frac{F_2 - 1}{G_1} + \frac{F_3 - 1}{G_1G_2} + \frac{F_4 - 1}{G_1G_2G_3} + \ldots$$

II - STATISTICS

Binomial or Bernouli distribution $p^n_{(x)} = C^n_x \, p^x \, q^{n-x}$

$\bar{x} = np \qquad \sigma^2 = npq$

Available shot noise power per hertz = $eIR/2$

e = charge on the electron (1.6×10^{-19} coulombs)

The Gaussian or normal distribution

$$p(x)\,dx = \frac{1}{\sqrt{2\pi}\,\sigma} \exp - [(x-\bar{x})^2/2\sigma^2]\,dx$$

The integral of the binomial mass function and the integral of the Gaussian density function between essentially the same limits is about the same if both distributions have the same means and standard deviations.

Rayleigh density function (linear detector output voltage)

$$p(r)\,dr = \frac{r}{\sigma^2} \exp - [\frac{r^2}{2\sigma^2}]\,dr$$

Rayleigh power density function (linear detector output power)

$$p(W)\,dW = \frac{R}{2\sigma^2} \exp - [\frac{WR}{2\sigma^2}]\,dW$$

Bayes Criterion for selecting decision threshold

$$\frac{f_1(x_t)}{f_0(x_t)} = \frac{(1-p)\,C_{10}}{p\,C_{01}}$$

Mean cycles to failure MCTF = $1/q$

variance $V(C) = \frac{p}{q^2}$

Mean time to failure MTTF = $1/\theta$

$$R(t_o) = e^{-\theta t_o}$$

III - SIGNAL PROCESSING AND DETECTION

$$(\frac{S}{N})_{out} = (\frac{S}{N})_{in} \sqrt{Bt} \quad \text{incoherent processing}$$

$$(S/N)_{out} = (S/N)_{in}\, 2Bt = E/N_o, \ \text{coherent processing}$$

approximate probability of false alarm or missed detection

$$P = \exp - \sqrt{\eta}$$

IV - ANTENNAS

$$\int_0^{2\pi} \int_0^{\pi} A_e(\theta,\phi) \sin \theta d\theta d\phi = \lambda^2$$

$$G(\theta,\phi) = 4\pi A_e(\theta,\phi)/\lambda^2$$

$$P_r = P_t G_t A_e/4\pi r^2$$

The approximate halfwidth of a circular antenna beam

$$\theta_1^2 \cong 4/G_{max} \text{ radians}$$

$$\theta_1 = \frac{2\lambda}{\pi d\sqrt{\rho_a}}$$

Approximate gain of a circular antenna

$$G_{max} = (2\pi a/\lambda)^2 \rho_a$$

V - PROPAGATION AND TRANSMISSION LINES

Maxwell's equations

$$\nabla \times H = \sigma E + j\omega\varepsilon E = (\sigma + j\omega\varepsilon) E$$

$$\nabla \times E = -j\omega \mu H$$

$$\nabla \cdot H = 0$$

$$\nabla \cdot E = \rho/\varepsilon$$

We can always get any equation containing ε back to its complete form by substituting

$$\left(\frac{\sigma}{j\omega} + \varepsilon \right) \text{ for } \varepsilon.$$

the wave equation

$$\nabla^2 H = -\omega^2 \mu\varepsilon H$$

$$\varepsilon_0 = 10^{-9}/36\pi \quad \mu_0 = 4\pi \times 10^{-7} \quad c = 3 \times 10^8 \text{ meters/sec.}$$

The maximum and minimum values of a solution of La Place's equation are always on the boundary

For a TEM wave

The velocity of propagation is $1/\sqrt{L_1C_1}$

$$\sqrt{L_1C_1} = \sqrt{\mu\varepsilon}$$

$$Z_0 = V(z)/I(z) = \sqrt{L_1/C_1}$$

For a Coaxial line

$$Z_0 = \sqrt{L_1/C_1} = \frac{1}{2\pi} \sqrt{\mu/\varepsilon} \ \ln (d_2/d_1)$$

$$\alpha = 4.34\ R/Z_0 \text{ dB/meter.}$$

For a plane wave

$$E_y/H_x = \sqrt{\mu/\varepsilon} = Z_0$$

Power flow $= H^2\sqrt{\mu/\varepsilon}$ or $E^2\sqrt{\varepsilon/\mu}$ per square meter

Skin depth $\delta = \dfrac{1}{\sqrt{\pi f \mu \sigma}}$

The total surface current which flows is equal to the value of the tangential magnetic field at the surface.

$$\frac{E_0}{J_y} = \frac{E_0}{H_0} = \sqrt{j\omega\mu/\sigma} = \frac{1+j}{\sigma\delta}$$

The losses in a conductor much thicker than the skin depth are the same as if a d.c. current of magnitude equal to the surface current or the tangential component of H at the surface, flowed uniformly in a slab of the material of thickness δ.

The effective r.f. resistance of the coaxial line per unit length, is is actually the resistance per unit length of a tube of thickness δ, the skin depth.

Group velocity
$$\frac{1}{v_g} = \frac{d}{df} \left(\frac{f}{v_p} \right)$$

Waveguide

$$V_p = \frac{1}{\sqrt{\mu\varepsilon}} \frac{1}{\sqrt{1 - f_c^2/f^2}}$$

Ionosphere

Plasma frequency $f_p = \frac{e}{2\pi} \sqrt{n/m\varepsilon_0}$

Treat as though $\varepsilon \rightarrow \varepsilon_0 \, [1 - (f_p/f)^2]$

$$V_p = \frac{1}{\sqrt{\mu\varepsilon}} \rightarrow \frac{1}{\sqrt{\mu\varepsilon_0}} \rightarrow \frac{1}{\sqrt{1 - f_p^2/f^2}}$$

$$\Gamma = \frac{Z_1 - Z_2}{Z_1 + Z_0}$$

VI - REFLECTION, REFRACTION AND SCATTERING

Snell's law $\sqrt{\mu_1\varepsilon_1} \cos \theta_1 = \sqrt{\mu_2\varepsilon_2} \cos \theta_2$

Brewster angle (E in the plane of incidence) no reflection when the direction of the reflected wave (if there were one) is perpendicular to the direction of the transmitted wave.

The radius of the circular path of the ray is $(dn/dz)^{-1}$ with the curvature being downward.

For average, or standard atmospheric conditions, the height of a nearly horizontal microwave beam or ray above the earth can be closely approximate by assuming that the ray is straight and the earth has 4/3 its actual radius of curvature.

VII - SOME SYSTEM CHARACTERISTICS

The noise reduction which occurs on demodulating a wide band FM signal is proportional to the square of the modulation index

Maximum Ideal Channel Capacity in bits per second $= B \log_2 (1 + n)$

Normal telephone practice uses 3.2 kHz bandwidth plus an 800Hz guardband

G/T is the antenna gain in dB minus the effective temperature of the receiver - antenna combination expressed in dB re 1°K

doppler shift (radar) $\Delta F = \frac{2v}{c} f_0$

APPENDIX B

TABLE OF SYMBOLS

This book combines a variety of subjects which are usually treated seperately. Each subject generally uses its own nomenclature with little or no coordination. Consequently there are many instances where the same symbol appears in different subjects with different meanings. To permit the reader to readily recognize those relationships with which he is already familiar, conventional notation is generally used in each discipline. This implies some duplication of symbols. In those cases where the same symbol has different meanings in different subject areas the meanings are identified by chapter or section.

The English letters are listed first, then the Greek letters.

A_e effective area of an antenna

B bandwidth

B_n noise bandwidth

C capacitance - Chapter I
number of calls - Section 2.8

C_1 capacitance per unit length

C_x^n binomial coefficient

CTF cycles to failure

CW continuous wave

c velocity of light

c_o velocity of light in free space

d detection index

d_1 inner diameter of coaxial line

d_2 outer diameter of coaxial line

dBW decibels relative to 1 watt

E energy

E electric field vector

E_r radial component of E

E_t energy per pulse

E_x x component of E

EIRP equivalent isotropic radiated power

e charge of the electron, 1.6×10^{-19} coulombs

F noise figure - except Section 2.7
number of failures - Section 2.7

FDM frequency division multiplex

f frequency

f_c cutoff frequency

f_m modulation frequency

f_p plasma frequency

G gain of amplifier or antenna

$G(\theta,\phi)$ gain of antenna in θ,ϕ direction

G/T antenna gain divided by temperature (dB)

H_{12} transfer admittance or impedance or voltage or current transfer function

H_x x component of magnetic field vector

H magnetic field vector

$I_B(<V_A>)$ the current at B due to a unit voltage at A

i.f. intermediate frequency

J_o surface current density

J_x x component of surface current density

j $\sqrt{-1}$

k Boltzmann's constant 1.38 x 10^{-23} - except Chapter V

arbitrary constant - Chapter V

L_1 inductance per unit length

L_a loss per unit of area

L_d loss per unit of length

ln natural log

MCTF mean cycles to failure

MTTF mean time to failure

m as a marginal notation see page 1 footnote

mass of electron - Chapter V

N noise power (watts)

N_{fa} number of false alarms per second

n number of trials - Chapter II

number of electrons per cubic meter - Chapter V

index of refraction - Chapter VI

P power - Chapter I, V, VII and VI except Section 6.2

probability - Chapter II

pressure - Section 6.2

$P_B(<V_A>)$ the power dissipated in resistor B due to unit voltage at A

$P^n_{(x)}$ probability of exactly x successes in n tries

PCM pulse code modulation

PPI plan position indicator

P_{fa} probability of a false alarm per opportunity

P_r received power

P_t transmitted power

P_{MD} probability of a missed detection

p probability of a sucess per trial - Chapter I, II

partial pressure of water vapor - Section 6.2

q 1-p

q_1 charge per unit length

R resistance - except Section 2.7, 7.4 and 7.6

reliability, fraction of componants not failed - Section 2.7

radius of the earth - Section 7.4

range - Section 7.6

r radius of satellite orbit - Section 7.4

radius of coaxial line-Chapter V

S signal power (watts)

S/N signal to noise ratio (η)

s signal voltage

T temperature, degrees Kelvin

T_e effective noise temperature

T_0 nominal room temperature (290°K)

T^1 transmission coefficient

TDM time division multiplex

TL terrestrial link

TTF time to failure

t time (seconds)

V voltage

V_g group velocity

V_p phase velocity

$V_B(<V_A>)$ the voltage at B due to unit voltage at A

v velocity

W power

x_T threshold or decision value of x

Z_0 characteristic impedance

z height - Section 6.2

α attenuation ratio (>1) not in dB - Chapter I
attenuation in dB/meter - Chapter V

Γ reflection coefficient

γ attenuation in nepers

Δf frequency deviation in frequency modulation

δ skin depth

ε dielectric constant

ε_0 dielectric constant of free space

θ probability of failure in small time interval t (exponential distribution) - Chapter II
angle - Chapter IV and VI

λ wavelength

μ permeability

η signal to noise power ratio (not in dB) - Chapter I, II, III
wave impedance - Chapter IV,V,VI

ρ_a aperture efficiency

σ standard deviation - Chapter II
conductivity - Chapter V
scattering cross section - Chapter VI, VII

Σ summation

$\phi(T)$ thermal equilibrium radiation

ω $2\pi f$

Ω solid angle

APPENDIX C

ALTERNATIVE PROOF OF RECIPROCAL PROPERTIES OF 2 PORT LOSSLESS NETWORKS

The proof to be given uses the 2 port scattering matrix.

$$\begin{bmatrix} S_{11} & S_{12} \\ S_{21} & S_{22} \end{bmatrix}$$

For a lossless junction this matrix is unitary, as shown for example in Ref.C1. This requires that

$$S_{11}S_{11}^* + S_{21}S_{21}^* = 1 \qquad (1)$$

$$S_{22}S_{22}^* + S_{12}S_{12}^* = 1 \qquad (2)$$

$$S_{11}S_{12}^* + S_{21}S_{22}^* = 0 \qquad (3)$$

The special case where $S_{22} = 0$ will be treated first. In this case equation 3 requires that either S_{11} or S_{12}^* be zero. However, if $S_{22} = 0$ and $S_{12}^* = 0$, equation 2 cannot hold. Therefore, $S_{11} = 0$ and the network is reciprocal in the sense discussed in Section 1.1 since Eq.1 and 2 then require that

$$|S_{21}| = 1 = |S_{12}|. \qquad (4)$$

For the more general case where $S_{22} \neq 0$, we first simplify the problem by moving the measurement planes on the input and output transmission lines so that S_{11} and S_{22} are real and positive. This does not affect the magnitude of any of the scattering coefficients but only their phases. Therefore, if $|S_{12}| = |S_{21}|$ using our moved measurement planes, it will be true for any measurement planes. In this case, from equation 2:

$$|S_{12}| = \sqrt{1 - (S_{22})^2} \qquad (5)$$

and from Eq.3, since $S_{22} \neq 0$,

$$|S_{21}| = \frac{S_{11}}{S_{22}} |S_{12}^*| = \frac{S_{11}}{S_{22}} |S_{12}|. \qquad (6)$$

---

*in this appendix the asterisk indicates complex conjugate.

From Eq.1 and 6

$$(S_{11})^2 + \frac{(S_{11})^2}{(S_{22})^2} \; |S_{12}|^2 \;=\; 1 \tag{7}$$

which becomes, on substituting Eq.5,

$$(S_{11})^2 + \frac{(S_{11})^2}{(S_{22})^2} [1 - (S_{22})^2] \;=\; 1$$

$$\frac{(S_{11})^2}{(S_{22})^2} \;=\; 1$$

$$S_{11} \;=\; S_{22} \tag{8}$$

Therefore, by Eq.1 and 2,

$$|S_{21}|^2 \;=\; 1 - (S_{11})^2 = 1 - (S_{22})^2 = |S_{11}|^2$$

and

$$|S_{21}| \;=\; |S_{12}| \tag{9}$$

which is the reciprocity condition demonstrated in Section 1.1 using the second law of thermodynamics.

A nonreciprocal gyrator device is frequently used as an isolator. Such a device is discussed in reference C1. It employs absorbing vanes in the waveguide configuration which makes it lossy. One might think that these vanes are not essential but, if fact, they are. The following brief discussion demonstrates this fact.

Consider a configuration in which the orientation of the waveguide is the same at the input and output sides of the gyrator. This configuration would involve a 45 degree rotation due to the ferite and 45 degree rotation due to a twist in the waveguide. For the waves travelling in one direction, these two rotations cancel so that the output wave is polarized in the same direction as the input wave and it passes on through the waveguide. In the other direction the two rotations add so that the output wave is at 90 degrees to the normal waveguide mode and will not pass. Thus we seem to have a lossless two port nonreciprocal device. However, let us consider what happens to this wave which cannot pass. It is, of course, reflected, returning through the gyrator in the reverse direction with compensating rotations so that it arrives back at its input still at 90 degrees to the normal mode of the guide. Again it cannot pass, so it is reflected once more, this time with a 90 degree phase shift which puts it in proper rotation for transmission out of the guide; so that after two reflections this wave will, in fact, continue on in the forward direction. Thus while the transient behavior may be somewhat different, CW measurements on such a device would indicate the same transmission coefficient in either direction provided the losses are truly negligible.

Reference C1 - Collins, R.E., Foundations of Microwave Engineering. McGraw-Hill, 1966

APPENDIX D

CALCULATION OF MEAN AND VARIANCE OF BINOMIAL DISTRIBUTION

The relation between the binomial distribution and the binomial theorum is expressed by

$$(p+q)^n = \sum_{x=0}^{x=n} C_x^n p^x q^{n-x} = \sum_{x=0}^{x=n} P_{(x)}^n \qquad (D.1)$$

The first equality in Eq.(D.1) is simply the binomial theorem.

The second equality comes from Eq.2.1 of Chapter 2. $P_{(x)}^n$ is the probability of exactly x successes in n trials. Since p + q = 1, we are assured by virtue of Eq.D.1 that the sum of all the probabilities is 1.

The mean value of x is defined as

$$\bar{x} = \sum_{x=1}^{x=n} x P_{(x)}^n = \sum_{x=1}^{x=n} x C_x^n p^x q^{n-x} \qquad (D.2)$$

Recognizing that each term in the summation of Eq.D.2 can be expressed as the derivative with respect to p of a term in the binomial expansion, we can write

$$\bar{x} = \sum_{x=1}^{x=n} p \frac{d}{dp} C_x^n p^x q^{n-x}$$

$$= p \frac{d}{dp} \sum_{x=1}^{x=n} C_x^n p^x q^{n-x}$$

$$= p \frac{d}{dp} (p+q)^n = np (p+q)^{n-1}$$

$$\bar{x} = np \qquad (D.3)$$

For the calculation of σ^2 or V(x) we need first a very general relationship which is quite independent of the type of statistics. By definition, the variance of x is

$$V(x) = \sum_x P(x) (x - \bar{x})^2$$

where P(x) = probability of the number x resulting from any experiment which can yield a range of numerical results.

Expanding the quadratic, we obtain

$$V(x) = \sum_x P(x)\,[x^2 - 2x\bar{x} + (\bar{x})^2]$$

$$= \sum_x x^2\,P(x) - 2\bar{x} \sum_x x\,P(x) + (\bar{x})^2 \sum_x P(x)$$

$$= \sum_x x^2\,P(x) - 2\overline{xx} + (\bar{x})^2$$

$$V(x) = \sum_x x^2 P(x) - (\bar{x})^2 \qquad (D.4)$$

Therefore for our case,

$$V(x) = \sum_{x=1}^{x=n} x^2\,p^n_{(x)} - n^2p^2$$

$$V(x) = \sum_{x=1}^{x=n} x^2\,C^n_x\,p^x q^{n-x} - n^2p^2 \qquad (D.5)$$

We now apply the same kind of trickery we used to obtain Eq.D.3 only in a slightly more sophisticated way.

$$\sum_{x=1}^{x=n} x^2\,C^n_x\,p^x\,q^{n-x} = p^2\,\frac{d^2}{dp^2}\,(p+q)^n + p\frac{d}{dp}\,(p+q)^n \qquad (D.6)$$

The proof of this equation is straightforward and will not be given here. The right hand side of Eq.D.6 reduces to

$$p^2\,n(n-1) + pn = n^2p^2 - p^2\,n + np = n^2p^2 + np(1-p)$$

and with this substitution Eq.D.5 reduces to

$$V(x) = \sigma^2 = npq \qquad (D.7)$$

APPENDIX E

GAUSSIAN DISTRIBUTION DERIVATIONS

In this appendix we will start with the Gaussian distribution in the form

$$P(x) \quad dx = \frac{1}{\sqrt{2\pi}b} \exp -[(x-a)^2/2b^2]dx \tag{E.1}$$

and demonstrate that the mean of the distribution is a and the standard deviation is b. It is these facts that permit us to write the distribution is its usual form (Eq.2.14 of Chapter 2).

First we must demonstrate that the integral over all probabilities is 1. This is difficult because Eq.E.1 is not integrable in closed form. For this purpose, consider the product

$$\int_{-\infty}^{+\infty} \frac{1}{\sqrt{2\pi}b} \exp [-(x-a)^2/2b^2] \, dx \int_{-\infty}^{+\infty} \frac{1}{\sqrt{2\pi}b} \exp [-(y-a)^2/2b^2] \, dy$$

$$= \frac{1}{2\pi b^2} \int_{-\infty}^{+\infty} \int_{-\infty}^{+\infty} \exp - [\frac{(x-a)^2 + (y-a)^2}{2b^2}]dx \, dy$$

where x and y are independent. If this double integral can be shown equal to 1, then each of the identical integrals must be 1, and we will have the proof we want. To show this we make a transformation to circular coordinates with origin at a, a. The integral then becomes

$$\frac{1}{2\pi b^2} \int_0^{2\pi} \int_0^{\infty} \exp [-r^2/2b^2] \, r \, dr \, d\theta$$

$$= \frac{1}{b^2} \int_0^{\infty} r \exp -[r^2/2b^2] \, dr$$

$$= [\exp - (r^2/2b^2)]_0^{\infty} = 1, \text{ q.e.d.}$$

Therefore,

$$\int_{-\infty}^{+\infty} \frac{1}{\sqrt{2\pi} \, b} \exp [-(x-a)^2/2b^2] = 1 \tag{E.2}$$

The mean of the distribution is given by

$$\bar{x} = \int_{-\infty}^{\infty} x\, P(x)\, dx = \frac{1}{\sqrt{2\pi}\, b} \int_{-\infty}^{\infty} x \exp\, [-(x-a)^2/2b^2]\, dx.$$

We make the transformation x-a = z.

$$\bar{x} = \frac{a}{\sqrt{2\pi}\, b} \int_{-\infty}^{\infty} \exp - \frac{z^2}{2b^2}\, dz + \frac{1}{\sqrt{2\pi}\, b} \int_{-\infty}^{\infty} z \exp - [z^2/2b^2]\, dz$$

$$= a - \frac{1}{\sqrt{2\pi}\, b}\, b^2\, [\exp - z^2/2b^2]_{-\infty}^{+\infty} = a - 0$$

Therefore, $\bar{x} = a$. (E.3)

The variance of the distribution is given by (see Eq.D.4 of appendix D),

$$\sigma^2 = \int_{-\infty}^{\infty} x^2\, P(x)\, dx - (\bar{x})^2$$

using the same transformation as before

$$(x-\bar{x}) = z, \quad x = z + \bar{x}, \quad x^2 = z^2 + 2z\bar{x} + (\bar{x})^2$$

$$\sigma^2 = \frac{1}{\sqrt{2\pi}\, b} \int_{-\infty}^{\infty} z^2 \exp - [z^2/2b^2]\, dz + \frac{2\bar{x}}{\sqrt{2\pi}\, b} \int_{-\infty}^{\infty} z \exp-[z^2/2b^2]\, dz$$

$$+ \frac{(\bar{x})^2}{\sqrt{2\pi}\, b} \int_{-\infty}^{\infty} \exp - [z^2/2b^2]\, dz - (\bar{x})^2$$

The last two integrals are readily recognized as 0 and 1, and the first integral can be handled by parts, letting u = z; dv = z exp - $[z^2/2b^2]$. The result is

$$\sigma^2 = \frac{b^2}{\sqrt{2\pi}\, b} [z \exp - (z^2/2b^2)]_{-\infty}^{\infty} + \frac{b}{\sqrt{2\pi}} \int_{-\infty}^{\infty} \exp - (z^2/2b^2)\, dz.$$

the first term gives zero, the second integral is $b\sqrt{2\pi}$, *thus*

$$\sigma = b$$

and the Gaussian distribution can be written in the form

$$P(x)dx = \frac{1}{\sqrt{2\pi}\sigma} \exp - \frac{(x-\bar{x})^2}{2\sigma^2} \qquad \text{(E.4)}$$

where σ and $\bar{x}$ have their usual meanings.

APPENDIX F

PROBLEMS

Chapter I

1. Show that a reciprocity theorum involving transfer impedence does not violate the second law of thermodynamics. (HINT - replace the zero impedence noise generator in series with the resistor by an equivalent infinite impedence generator in parallel with it.)

2. Show that the average noise power stored in an inductance is kT/2

3. Find the noise figure of:
   (a) A 3 dB attenuator at temperature 290
   (b) A 10 dB attenuator at temperature 360
   (c) A 30 dB attenuator at temperature 410
   (d) A cascade of a 10 dB amplifier F = 3 dB followed by a 20 dB amplifier F = 10 dB (same bandwidth)
   (e) A cascade of a 10 dB amplifier F = 10 followed by a 3 dB amplifier F = 3 dB (same bandwidth)
   (f) A mixer, loss 3 dB F = 6 dB, followed by a 20 dB amplifier F=10 dB.

4. Find the noise output (in watts) of an amplifier of bandwidth 10 MHz gain 20 dB and F = 3 dB, fed from a mixer of loss 6 dB, F = 10 dB which in turn is fed from a matched resistor at 5°K.

5. What is the effective noise temperature of the mixer and amplifier combination in problem 4?

6. At the output of the amplifier in 4 above, I add a second amplifier with effective noise temperature 2000°K bandwidth 2 MHz and gain 10 dB. What is the new noise output?

7. A large radar dish looks at a sky temperature of 6°K through an atmosphere at 200°K which attenuates 1 dB. It employs a cryogenic amplifier with T_e = 5°K, bandwidth 10 MHz and gain 10 dB, followed by a conventional amplifier of gain 20 dB, F = 3 dB bandwidth 2 MHz:
   (a) What size signal do I need at the input to the cryogenic amplifier to have S/N = 1 at the output?
   (b) How would a) change if the atmospheric attenuation were 3 dB?
   (c) How would a) change if the atmospheric attenuation were 0 dB?
   (d) How would a) change if the cryogenic amplifier bandwidth were 1 MHz?

Chapter II

1. Calculate the expected number of 7's in 4 throws of a pair of dice by summing over the probabilities of 1, 2, 3, and 4 sevens.

2. Of families with two children, at least one of which is a boy, what is the probability that the other is a boy? (HINT - look out for statistical independence.)

3. Prove Equation 2.7.

4. The limit of the binomial distribution for the case of large n and small p is the Poisson Distribution

$$P(x) = \exp(-m)\, m^x/x! \quad \text{where m is the mean}$$

Suppose a nuclear counter is counting at an average rate of 1 per second in what fraction of the 1 second intervals do I expect to find 0, 2, 3, and 5 counts? If I look at 5 second intervals, in what fraction of them do I expect to find 0 counts, 5 counts? In what fraction of the 5 second intervals do I expect to find 7 or more counts. Estimate this last result from the normal distribution using the DeMoivre LaPlace limit theorum.

5. In problem 4, in what fraction of the 1 second intervals do I expect to find 4 or more counts? Use the Poisson and normal distributions to get the answer and compare results.

6. What is the shot noise power per megahertz delivered to a 500 ohm load by a temperature limited diode carrying 5 milliamperes?

7. A message is conveyed by a 5 bit code (each bit is a 1 or a 0). I can tolerate 1 error in 10^4 messages.
   a. What is my tolerable probability of error per bit?
   b. If the errors are to be equally divided between false alarms and missed detections, what is my tolerable false alarm probability per bit?
   c. What is the approximate ratio of detection threshold to noise which I need to achieve that false alarm rate?

8. A digital communication system is used to transmit signals as zeros and ones. If a signal is present it is scored as a one. On the average only 1/5 of the signal intervals are filled. The penalty for missing a one is twice the penalty for missing a zero. At what likelihood ratio should my receiver threshold be set according to the Bayes criterion?

9. a. How many units must I test to assure 90% reliability with 80% confidence if I get no failures among my samples?
   b. Suppose I get 1 failure in 10 tests, with what confidence can I state that the reliability is 80%? 70%?
   c. What would the answer to (b) be if I got 2 failures in 20 tests?

10. The mean lifetime of 5 transistors is 2 years. With what confidence can I state that a transistor of that batch will operate for 1 year assuming an exponential distribution?

11. Suppose I have a piece of equipment with two transistors of the batch described in problem 10. The equipment will fail if either fails. With what confidence can I state that the equipment will operate for a year?

12. Repeat problem 11 if the equipment is designed with redundancy so that it will only fail if both transistors fail.

13. a. How large a time interval do I need to measure a microampere to an accuracy of 1% with 99% confidence under shot noise conditions? (Use the DeMoivre-LaPlace limit theorum)
    b. What if it were 10^{-12} amperes?

14. The peak load on communication circuits occurs at 3 PM. At that time the mean number of calls over a particular link is 20, and over a main trunk link is 350. What capacity do I need in each link to achieve 99% probability of adequate capacity? What will the average % utilization of each link capacity be at that hour if I design for 99% probability of adequate capacity?

<u>Chapter III</u>

1. I intend to transmit 10^6 bits of information per second over a 1 MHz bandwidth. The signal to noise ratio at the input to the threshold device is 10 dB. Using Eq.3.12, approximately what error rate must I

tolerate?

2. In problem 1, I reduce the information rate to 2 x 10^5 bits of information per second and use incoherent signal processing. What is the new error rate?

3. In problem 2 I use coherent processing. What is the error rate?

4. Repeat problems 1, 2 and 3 using the ROC curves with the missed detection probability equal to the false alarm probability.

5. Repeat problem 3 using the ROC curves with the probability of a missed detection equal to 2 x the probability of a false alarm.

6. Suppose a false alarm was 100 times as costly as a missed detection and the apriori probabilities were the same. Find the Bayes Criterion solution to problem 3 in terms of false alarm and missed detection rates.

7. The noise into a receiver is thermal noise at a temperature of 400°K. The receiver consists of a mixer of 3 dB loss and noise figure 6 dB, followed by an i.f. amplifier of noise figure 6 dB and bandwidth 2 megahertz, followed by noiseless detection, averaging, and threshold circuits. The data rate is 10^5 bits per second with half the bits being zero and half ones. I use incoherent processing gain in an optimum way and I can tolerate only 1 error per hundred bits.
   a. What is the overall effective noise temperature of the receiver?
   b. What detection index do I need to achieve the required error rate?
   c. What signal level is required at the receiver input?

<u>Chapter IV</u>

1. A satellite at a distance of 22,000 miles from the earth transmits to the earth with an antenna having a 20 dB gain. What is the approximate radius of the circle on the earth that it illuminates?

2. Repeat problem 1 for a antenna with a gain of 60 dB.

3. A radio astronomy telescope has a main beam efficiency of 80% (80% of its radiated power is transmitted in the main beam). It is used as a receiver with its main beam looking at a sky temperature of 5°K, and its side lobes looking at the ground with temperature 290°K. What is the effective noise temperature of the antenna?

4. In problem 3 I add a metal mesh on the ground with reflectivity 90%. The antenna sees the sky reflected in the mesh except for 10% loss of energy on reflection (treat as an attenuator). What is the new effective noise temperature of the antenna?

5. What is the effective temperature of a lossless omnidirectional receiving antenna in a satellite at a height of 22,000 miles? (The earth is 8,000 miles in diameter and at a temperature of 200°K; space temperature is 5°K.) Assume the antenna cannot see the sun.

6. How much will the sun affect the effective temperature of the antenna in problem 5 if it is at 6,000°K and subtends an angle of 32 minutes?

7. How much power must I transmit from earth through a 50 dB gain antenna to achieve a signal to noise ratio of 20 dB at the output of the omnidirectional antenna of problem 5 if the frequency is 6 GHz and the bandwidth is 10 MHz?

8. The National Science Foundation is funding a project to build a radio astronomy telescope 25 miles long. What angular resolution will it have directly overhead at 10 gigahertz? At 1 gigahertz? Assume that the distribution of antennas along the 25 mile length is continuous.

9. I have a cryogenic receiver with a bandwidth of 10 KHz at a frequency of 3 GHz and an effective noise temperature of 10°K, connected to an antenna with an effective temperature of 5°K, receiving signals from a

100 watt omnidirectional transmitter orbiting Mars at a distance of 50 million miles from earth. How large an earth antenna do I need to have a signal to noise ratio of 10 dB at the receiver output?

10. If I use coherent processing in problem 9 and hold the bit rate to 10 per second, how large an earth antenna do I need?

Chapter V

1. Show from the basic definition of inductance (e = L di/dt) and Maxwell's equations that the inductance per unit length of coaxial line is given by

$$L_1 = \frac{\mu}{2\pi} \ln (d_2/d_1)$$

2. Find the characteristic impedance of a coaxial line of inner diameter 1 Cm and outer diameter 2 Cm filled with lossless material of dielectric constant 2 times that of space.

3. Find the attenuation in dB per kilometer of such a line if its conductors are copper (σ = 5.8 x 10^7 mhos/meter) at a frequency of

   a) 1 megahertz
   b) 10 megahertz
   c) 100 megahertz

4. How would the answer to 3 change if

   a) the line were air filled
   b) the line were filled with the same lossless material but were twice as large

5. For a fixed outer diameter of coaxial line filled with lossless material of dielectric constant 3 times that of air, what ratio of inner to outer diameter gives the minimum attenuation in dB/meter? What is its characteristic impedance?

6. Find the wavelength, the wave impedance (the ratio of E to H), and the attenuation in dB per meter of a plane wave in sea water (σ = 4 mhos/meter) at a frequency of

   a) 10 kilohertz
   b) 100 kilohertz
   c) 1 megahertz

7. An air filled copper waveguide 3 Cm wide and 1 Cm high carries the TE_{10} mode.

   a) What is its attenuation in dB/meter at 10,000 megahertz?
   b) What is its attenuation in dB/meter at 6000 megahertz?
   c) How would the answer to a) change if it were 1/2 Cm high?

8. The waveguide of problem 7 at a temperature of 290°K is part of the feed system of an antenna receiving signals from a satellite at 6000 megahertz. It is 2 meters long. The receiver noise temperature is 20°K, and the effective temperature seen by the antenna is 5°K.

   a) What is the effective noise temperature of the antenna-waveguide-receiver system if the waveguide is 1 Cm high?
   b) What would it be if the waveguide is 1/2 Cm high?
   c) What would it be if the waveguide were lossless?

9. The ionosphere has an electron density, at some particular time, of 10^{10} electrons per cubic meter. Find

   a) The plasma frequency
   b) The plane wave impedance at a frequency of 3 megahertz
   c) The reflection coefficient at normal incidence for a plane wave entering the ionosphere from air at 3 megahertz, assuming the ionosphere begins abruptly.

10. If the electron density in interstellar space were 10 electrons per

cubic meter.

a) Find the plasma frequency f_p.

b) Find the difference between the phase velocity of propagation in the plasma and what it would be in a true vacuum (a small quantity) as a function of frequency for $f >> f_p$. (A first order approximation is adequate and easier to calculate)

c) Find the group velocity and show the first order deviation from c.

d) Find the percent difference in group velocity between f = 80 megahertz and f = 120 megahertz.

e) What would be the difference in arrival time of two signals at these two frequencies that started toward earth at the same time from a source 10 light years away.

Chapter VI

1. What is the critical angle for total internal reflection in a material with

   a) $\varepsilon' = 2\varepsilon_0 \quad \mu' = \mu_0$

   b) $\varepsilon' = \varepsilon_0 \quad \mu' = 2\mu_0$

2. What is the critical angle for total internal reflection in sea water (σ = 5 mhos/meter, $\varepsilon = 81\ \varepsilon_0$) at 1 MHz?

3. a) Find the Brewster angle for incidence from air into distilled water $\varepsilon' = 81\ \varepsilon_0$, $\sigma = 0$, at 1000 megahertz. Show that the direction of the transmitted wave and the direction in which the reflected wave would go if there were one, are mutually perpendicular.

   b) At this same angle, what will the reflected amplitude be if it is lake water instead of distilled water ($\sigma = 5 \times 10^{-3}$)

4. Find the reflection coefficient of dry earth with $\varepsilon' = 4\varepsilon_0$, $\sigma = 0$ with E in the plane of incidence at grazing angles of

   a) 5°

   b) 10°

   c) 45°

   d) 90°

5. Repeat problem 4 with E perpendicular to the plane of incidence.

6. Find the radius of curvature of a nearly horizontal microwave path at sea level at 260° Kelvin with pressure decreasing at the rate of 3×10^{-5} atmosphere per foot of altitude.

   a) If the air is completely dry

   b) If the partial pressure of water vapor is constant at 2%

   c) If the partial pressure of water vapor is 2% at sea level and decreases uniformly to 1% at 5000 feet.

7. a) Find the radius of curvature of a nearly horizontal microwave path at sea level with pressure decreasing 3×10^{-5} atmospheres per foot of altitude and temperature 240° Kelvin at sea level decreasing 10° per thousand feet of altitude. The air is dry.

   b) Calculate the equivalent earth curvature radius for this condition.

8. Under standard atmospheric conditions with a 40 mile path,

   a) What is the total angle through which the ray path turns?

   b) How far in feet does a ray path deviate from a straight line continuation of its original path?

   c) If the ray path were straight, the earth were perfectly flat and the transmitting tower were 100 ft high, how high would the receiving tower 40 miles away have to be to receive the ray that just grazed the earth?

d) Repeat c) with the standard atmosphere refraction instead of the straight ray path.

9. Two omnidirectional antennas are 10 kilometers apart at ground level. An aeroplane with scattering cross section 10 square meters is midway between them at a height of 2000 meters. Calculate the ratio of the direct signal voltage to the scattered signal voltage at the receiving antenna.

10. Repeat problem 9 with the aeroplane 1 Kilometer from one of the antennas in the direction of the other, instead of midway between them.

11. A microwave link at 10,000 megahertz uses a large reflecting surface mirror to reflect a signal around a hillfrom the transmitting antenna to the receiving antenna. The surface is 3 meters square and oriented so as to reflect the signal directly to the receiving antenna. The distance between the antennas is 10 Km, and the reflector is midway between the antenna and 2 Km off the direct path. Assuming that the receiving and transmitting antennas are omnidirectional, what is the ratio of the received signal to what it would be if the hill and reflector were not there?

Chapter VII

1. An earth station receiving antenna of the INTELSAT system is 100 feet in diameter and operates at 4 GHz. If the sky background temperature is 5°K and G/T is 40 dB, by how many dB is the signal to noise ratio at the receiver output changed when the sun transits the field of view of the antenna? Assume that the sun subtends an angle of 30 minutes and is at a temperature of 6000°K.

2. An airplane with a scattering cross section of 20 square meters flies across the line of sight between a satellite and an earth antenna at a height of 1000 meters. Assume that it does not significantly interrupt the direct signal between the two antennas but adds a scattered path. How much will the received signal power be reduced if the voltage in this reflected signal arrives out of phase with the direct signal voltage

INDEX